CONTENTS

MANUFACTURING AUTOMATION

Metal cutting is one of the most widely used methods of producing the final shape of manufactured products. The technology involved in metal cutting operations has advanced considerably in recent years along with developments in materials, computers, and sensors.

This book treats the scientific principles of metal cutting and their practical application to solving problems encountered in manufacturing. The subjects of mathematics, physics, computers, software, and instrumentation are discussed as integration tools in analyzing or designing machine tools and manufacturing processes.

The book begins with the fundamentals of metal cutting mechanics. Basic principles of vibration and experimental modal analysis are applied to solving problems on the shop floor. A special feature is the in-depth coverage of chatter vibrations, a problem experienced daily by practicing manufacturing engineers. The essential topics of programming, design, and automation of CNC (computer numerical control) machine tools; NC (numerical control) programming; and CAD/CAM technology are fully discussed. The text also covers the selection of drive actuators, feedback sensors, modeling and analysis of feed drives, the design of real time trajectory generation and interpolation algorithms, and CNC-oriented error analysis in detail. Each chapter includes examples drawn from industry, design projects, and homework problems.

Advanced undergraduate and graduate students, as well as practicing engineers, will find this book a clear and thorough way to learn the engineering principles of metal cutting mechanics, machine tool vibrations, CNC system design, sensor-assisted machining, and CAD/CAM technology.

Yusuf Altintas is Professor of Mechanical Engineering and Director of the Manufacturing Automation Laboratory at the University of British Columbia.

MANUFACTURING AUTOMATION

METAL CUTTING MECHANICS, MACHINE TOOL VIBRATIONS, AND CNC DESIGN

YUSUF ALTINTAS

University of British Columbia

CAMBRIDGE
UNIVERSITY PRESS

PUBLISHED BY THE PRESS SYNDICATE OF THE UNIVERSITY OF CAMBRIDGE
The Pitt Building, Trumpington Street, Cambridge, United Kingdom

CAMBRIDGE UNIVERSITY PRESS
The Edinburgh Building, Cambridge CB2 2RU, UK   http://www.cup.cam.ac.uk
40 West 20th Street, New York, NY 10011-4211, USA   http://www.cup.org
10 Stamford Road, Oakleigh, Melbourne 3166, Australia
Ruiz de Alarcón 13, 28014 Madrid, Spain

First published 2000

Printed in the United States of America

Typeface Century Schoolbook 10/13 pt. and Univers   *System* LaTeX 2_ε [TB]

A catalog record for this book is available from the British Library.

Library of Congress Cataloging in Publication Data
Altintas, Yusuf, 1954–
     Manufacturing automation : metal cutting mechanics, machine tool
vibrations, and CNC design / Yusuf Altintas.
          p.   cm.
     ISBN 0-521-65029-1 (hc.). – ISBN 0-521-65973-6 (pbk.)
     1. Machining – Automation.   2. Machine-tools – Vibration.
3. Machine-tools – Numerical control.   I. Title.
     TJ1185.5.A48   2000
     671.3′5 – dc21
                                                                99-30935
                                                                     CIP

ISBN 0 521 65029 1 hardback
ISBN 0 521 65973 6 paperback

**APPENDIX B: OFF-LINE AND ON-LINE PARAMETER ESTIMATION
WITH LEAST SQUARES**   273

   B.1 Off-Line Least Squares Estimation   273
   B.2 Recursive Parameter Estimation Algorithm   275

Bibliography   277
Index   283

PREFACE

Metal cutting is one of the most widely used manufacturing processes to produce the final shape of products, and its technology continues to advance in parallel with developments in materials, computers, and sensors. A blank is converted into a final product by cutting extra material away by turning, drilling, milling, broaching, boring, and grinding operations conducted on Computer Numerically Controlled (CNC) machine tools. The purpose of this book is to help students and engineers understand the scientific principles of metal cutting technology and the practical application of engineering principles to solving problems encountered in manufacturing shops. The book reflects the author's industrial and research experience, as well as his manufacturing engineering philosophy.

Engineers can learn best by being shown how to apply the fundamentals of physics to actual machines and processes that they can feel and visualize. Mathematics, physics, computers, software packages, and instrumentation then become useful integration tools in analyzing or designing machine tools and machining processes.

Metal cutting operations take place between a cutting tool and workpiece material mounted on a machine tool. The motion of the machine tool is controlled by its CNC unit, and the Numerically Controlled (NC) commands to CNC are generated on computer-aided design/computer-aided manufacturing (CAD/CAM) systems. The productivity and accuracy of the metal removal operation depend on the preparation of NC programs, planning of machining process parameters and cutting conditions, cutter geometry, work and tool materials, machine tool rigidity, and performance of the CNC unit. Manufacturing engineers who are involved in machining and machine tool technology must be familiar with each of these topics. It is equally important to link them and to be able to apply them in an interdisciplinary fashion to solve machining problems.

The beginning chapters of this book provide detailed mathematical models of metal cutting, milling, and turning operations. The macromechanics of cutting, which is applicable to solving problems on the shop floor and in machine tool design, is emphasized. Although work and tool material design – the micromechanics of cutting – is not covered in the text, basic principles of machinability, tool wear mechanisms, and chipping are briefly introduced to provide a complete picture. The design of machine tools requires knowledge of structures, mechanics of solids, vibrations, and kinematics, subjects that are covered in dedicated mechanical engineering texts. This text builds on that knowledge, applying the principles of vibration and experimental modal analysis to machine tools and metal cutting. Mathematical methods are simplified so that they can be easily used to solve machining vibration problems. Chatter vibrations in machining are treated in depth in this text since the problem is experienced daily by practicing manufacturing engineers.

The last three chapters of the book are dedicated to programming, design, and automation of CNC machine tools. Numerically Controlled programming and CAD/CAM technology are briefly covered, but with sufficient explanation so that the reader can start programming and using CNC machine tools. The selection of drive actuators, feedback sensors, modeling and analysis of feed drives, the design of real time trajectory generation and interpolation algorithms, and CNC oriented error analysis are presented in more detail than can be found in other texts. Open CNC design philosophy and improvement of accuracy and productivity by adding sensors and algorithms to CNC machine tools are also fully covered.

Students learn best from dealing with real manufacturing problems. The contents of this book are based on experimentally proven engineering principles that are widely used in applied research laboratories and industry. The examples and problems presented in each chapter originate from the research and industrial problems solved by the author and his graduate students. Interdisciplinary problems are posed as industrial projects so that readers can apply all the necessary techniques simultaneously. They solve the basic metal cutting mechanics problem first, followed by milling mechanics, static deflection of end mills and corresponding surface-form error modeling, and vibration model of the end mill and chatter stability, respectively. For example, the chain of knowledge is exercised in solving problems associated with milling of an aircraft structure, a project that originated from industry. Similarly, in another project, the reader is guided step-by-step through the programming, real time modeling, and control of a CNC machine tool. Since all the projects were tried in the author's laboratory, a number of teaching and research setups are provided in the book to aid instructors.

The book is intended as a text for senior undergraduate and graduate students and practicing manufacturing engineers who wish to learn the engineering principles of metal cutting, machine tool vibration, experimental modal analysis, NC programming and CAD/CAM technology, CNC system design, and sensor-based intelligent machining. The book can also be used by researchers who wish to study metal cutting mechanics, dynamics, and CNC and sensor-based machining.

Acknowledgments

The contents of each chapter mostly originated from the author's own engineering, research, and teaching experience. Except for the CAM technology (Chapter Four), each chapter is based on a number of graduate student theses supervised at the Manufacturing Automation Laboratory of the author at the University of British Columbia.

The studies of former graduate students Dr. E. Budak, Dr. A. Spence, and Haikun Ren have contributed to Chapter Two, which deals with metal cutting. Dr. E. Schamoto contributed to the theory of shear angle prediction in oblique cutting operations. The theses of E. Budak, S. Engin, D. Montgomery, P. Chan, P. Lee, and M. Campomanes were very helpful in writing Chapter Three, where the machine tool deformations and vibrations are presented. The theses of electronics engineers J. Peng and A. Nick Newell and mechanical engineers A. J. Lane and K. Erkorkmaz were helpful in writing Chapter Five, which presents the principles of CNC design. The graduate research theses of N. A. Erol and K. Munasinghe were instrumental in forming the final chapter, which deals with sensor-fused machining and open CNC system design. The author acknowledges the contributions of all of his former and present graduate students to the accumulation and dissemination of knowledge in machining, machine tools, CNC design, and machining process monitoring and control. Dr. S. Atadan, who has been a close friend for many years, assisted in preparing the overview of Laplace and z transforms presented in the appendix. Dr. I. Lazoglu assisted in proofreading the text.

Several machinists, engineers, and professors contributed to the author's manufacturing engineering experience. The author received significant practical training from the machine tool design engineers, technologists, and machinists at M.K.E. Top Otomotiv Factory in Kirikkale, Turkey; the machinists at Pratt & Whitney Canada in Montreal; and the process planners of Canadian Institute of Metal Working in Hamilton. The basic engineering education with rich machine design and analytical content from Istanbul Technical University, CAD/CAM education from the University of New Brunswick, and machine tool engineering background from McMaster University were most valuable in overall development of the author's manufacturing engineering and research skills. Prof. G. Ulsoy, Prof. Y. Koren, Prof. G. Pritschow, Prof. U. Heisel, Prof. T. Moriwaki, and Prof. T. Altan provided strong personal friendship, research partnership, and encouraged the author in completing this text. The author was and has been most influenced by the research and machine tool engineering style and philosophy of his former supervisor and mentor Prof. J. Tlusty.

I acknowledge the valuable editorial support provided by the Cambridge University Press editor, Florence Padgett. Project managers Andrew Wilson and Michie Shaw of TechBooks handled the final production of the manuscript with great care and professionalism. I thank those in Cambridge University Press and TechBooks for their cooperation and assistance.

Machine tool and metal cutting engineering is a multidisciplinary area that demands knowledge from various fields if one is to be an effective manufacturing engineer and researcher. This requires significant work and effort, which cannot be accomplished without the sacrifices of close family members. The author's wife Nesrin, daughter Cagla, and son Hasan bore with the author, who missed an endless number of weekends and family days in establishing the Manufacturing Automation Research Laboratory at UBC, where the book has been developed over twelve years. The author's father Hasan and mother Hatice, who are hardworking, honest, and warm country folks living in the little town Bekili (Denizli, Turkey), were more than a role model but a source of support for the author. The author owes the completion of the book to those who supported him throughout his career.

CHAPTER ONE

INTRODUCTION

The area of machine tools, metal cutting, computer numerically controlled (CNC), computer-aided manufacturing (CAM), and sensor-assisted intelligent machining is quite wide, and each requires an academic and engineering experience to appreciate a manufacturing operation that uses all of them in an integrated fashion.

Although it is impossible to be an expert in all these subjects, a manufacturing engineer must be familiar with the engineering fundamentals for the precision and economical manufacturing of a part. This book emphasizes only the fundamentals of metal cutting mechanics, static and dynamic deformations, design principles of CNC, sensor-assisted machining, and technology of programming CNC machines. Although industrial engineering aspects, such as the scheduling of parts and economics, also play an important role in cost effective manufacturing, the book focuses only on the physical principles of machining that reflect the engineering, research and teaching experience of the author.

The book is organized as follows:

Chapter Two covers the fundamentals of metal cutting mechanics. The mechanics of two-dimensional orthogonal cutting is introduced first. The laws of fundamental chip formation and friction between the rake and flank faces of a tool during cutting are explained. The relationships among the workpiece material properties, tool geometry, and cutting conditions are presented. Identification of shear angle, average friction coefficient between the tool's rake face and moving chip, and yield shear stress during machining is explained. The oblique geometry of practical cutting tools used in machining is introduced. The mechanics of oblique cutting for three-dimensional practical tools are explained, and methods in predicting the cutting forces in all directions are presented using the laws of oblique cutting mechanics. Mechanics of turning, milling and drilling, which constitute the majority of machining operations in the manufacturing industry, are presented. Algorithms for predicting the milling forces in three Cartesian coordinates are derived and illustrated with sample experimental results. Efficient force prediction algorithms for widely used helical end mills are presented. The chapter also briefly discusses the modes and causes of tool wear and breakage, which are important in evaluating the machinability of parts.

Chapter Three deals with static deformations and vibrations during machining. The static deformations occur due to elastic deflections of both parts and machine during machining. When the static deformation is beyond the tolerance limit, the part is scrapped. Sample formulations are provided to predict the magnitude and location of static deformations in bar turning and end milling. The methods can be extrapolated to other machining operations such as grinding and drilling. One of the most common problems in machining originates from dynamic deformations (i.e., relative structural vibrations between the tool and workpiece). The most common vibrations are due to self-excited chatter vibrations, which grow until the tool jumps out of the cutting zone or breaks due to exponentially growing dynamic displacements between the tool and workpiece. In order to understand the machine tool vibrations, the fundamental principles of single- and multi–degree-of-freedom vibration theory are summarized first. Since the machine tool chatter is mainly investigated by analyzing experimental data, the fundamentals of the experimental modal analysis techniques are presented. The modal analysis technique allows the engineer to represent a complex machine tool or workpiece structure by a set of commonly used mathematical expressions that engineers can understand. The technique not only allows one to analyze the chatter vibrations, but it gives a clear message to the machine tool engineer about the structural source of the vibrations during machining, which leads to the improved design. After providing basics of vibration engineering, the chapter presents the theory of chatter vibrations in both orthogonal machining and complex milling operations. The mathematical model of regenerative vibrations, which occur in subsequent tool passes during machining, is presented. Methods of determining chatter vibration–free axial depths of cuts and spindle speeds in orthogonal cutting operations are presented. A special technique that allows the engineer to predict chatter-free conditions in complex milling operations is introduced. The technique is explained with the aid of results obtained from simulation and machining tests. The engineer is presented with methods that increase the machining productivity by avoiding chatter vibrations.

Chapter Four introduces the CNC technology and its principles of operation and programming. First, standard NC commands accepted by all CNC machine tools are summarized. These include the format of an NC code accepted by the CNC of the machine tool, motion commands such as linear and circular contouring along a tool path, miscellaneous commands such as spindle and coolant control, and automatic cycles. Later, computer-aided manufacturing (CAM) is introduced by explaining the difficulties in generating NC codes manually. First, the APT (Automatically Programmed Tooling) programming language is introduced. The structure of parametric definition of part geometry, tool geometry, and tool motions is explained, along with sample APT language commands. The general output of the APT, which is called CL (cutter location) data, is standard and acceptable by all CNC machine tools when it is processed by a postprocessor. The postprocessor converts the machine tool–independent CL data into machine tool–specific manual NC commands automatically. Later, computer-aided design (CAD), used both in designing the parts and generating tool paths by interactive graphical techniques, is explained. The chapter focuses on teaching the basics and the use of CAM technology, but not the theory of computer graphics, which is not within the main mandate of a manufacturing engineer.

Engineers who know how to use and program CNC machine tools must familiarize themselves with the design and internal operational principles of CNC. Chapter Five describes the fundamentals of CNC design, starting with the selection of drive motors and servo amplifiers. Mathematical modeling of feed servo drives is presented in detail. Transfer functions of mechanical drive inertia and friction, servo motor, amplifier, and velocity and position feedback sensors are explained with their practical interpretations. Transforming continuous-time domain models of the physical system into discrete computer time domain models is explained. Design and tuning procedures for digital control of feed drives are presented with real life examples. The CNC design includes real-time tool path generation with linear, circular, and spline interpolations coupled with the scheduling of acceleration, deceleration, and vector feed speed control along the tool paths. The chapter is complemented with the design of electro-hydraulic machine tool drives in order to show that the CNC design principles are general and can be applied to any mechanical system regardless of the actuators.

The recent trend in machining is to add intelligence to the machine tools and CNC, as discussed in Chapter Six. Sensors, which can measure the forces, vibrations, temperature, and sound during machining, are installed on the machine tools. Mathematical models, which correlate the relationship between the measured sensor signals and state of machining, are formed. The mathematical models are coded into real-time algorithms, which monitor the machining process and send commands to CNC for corrective actions. The chapter presents an Intelligent Machining and Open Real Time CNC Operating System that allows modular design and integration of sensor-fused application algorithms into the machine tool. The chapter includes simple but fundamental machining process control algorithms along with their theoretical foundations. Adaptive control of cutting forces, in-process monitoring of tool failure, and chatter detection algorithms are presented with their experimental validation and engineering application.

Sample problem sets are included at the end of each chapter. The problems mostly originated from the actual design, application, and experiments conducted at the author's manufacturing automation research laboratory; hence they are designed to give a realistic feeling for engineering students. Since the book contains multiple engineering disciplines applied to machine tool engineering problems in an integrated fashion, most of the basic mechanical engineering concepts are assumed to be understood by the readers. However, the basic principles of Laplace and z transforms, as well as least square–based identification techniques, are provided in the appendix.

MECHANICS OF METAL CUTTING

2.1 INTRODUCTION

The final shapes of most mechanical parts are obtained by machining operations. Bulk deformation processes, such as forging and rolling, and casting processes are mostly followed by a series of metal removing operations in order to achieve parts with desired shapes, dimensions, and surface finish quality. The machining operations can be classified under two major categories: cutting and grinding processes. The cutting operations are used to remove material from the blank. The subsequent grinding operations provide a good surface finish and precision dimensions to the part. The most common cutting operations are *turning*, *milling*, and *drilling* followed by special operations such as *boring, broaching, hobing, shaping*, and *form cutting*. However, all metal cutting operations share the same principles of mechanics, but their geometry and kinematics may differ from each other. The mechanics of cutting and the specific analysis for a variety of machining operations and tool geometries are not widely covered in this text. Instead, a brief introduction to the fundamentals of cutting mechanics and a comprehensive discussion of the mechanics of milling operations are presented. Readers are referred to established metal cutting texts authored by Armarego and Brown [2], Shaw [3], and Oxley [4] for detailed treatment of the machining processes.

2.2 MECHANICS OF ORTHOGONAL CUTTING

Although the most common cutting operations are three dimensional and geometrically complex, the simple case of two-dimensional orthogonal cutting is used to explain the general mechanics of metal removal. In orthogonal cutting, the material is removed by a cutting edge that is perpendicular to the direction of relative tool–workpiece motion. The mechanics of more complex three-dimensional oblique cutting operations are usually evaluated by geometrical and kinematic transformation models applied to the orthogonal cutting process. Schematic representations of orthogonal and oblique cutting processes are shown in Figure 2.1. The orthogonal cutting resembles a shaping process with a straight tool whose cutting edge is perpendicular to the cutting velocity (V). A metal chip with a width of cut (b) and depth of cut (h) is sheared away from the workpiece. In orthogonal cutting, the cutting is assumed to be uniform along the cutting edge; therefore it is a two-dimensional plane strain deformation process without side spreading of the material. Hence, the cutting forces are exerted only in the directions of velocity and uncut chip thickness, which are called tangential (F_t) and feed forces (F_f). However, in oblique cutting, the cutting edge is oriented with an inclination angle (i) and the additional third force acts in the radial direction (F_r).

There are three deformation zones in the cutting process as shown in the cross-sectional view of the orthogonal cutting (see Fig. 2.2). As the edge of the tool penetrates into the workpiece, the material ahead of the tool is sheared over the primary shear zone to form a chip. The sheared material, the chip, partially deforms and moves along the rake face of the tool, which is called the secondary deformation zone. The friction area, where the flank of the tool rubs the newly machined surface, is called the tertiary zone. The chip initially sticks to the rake face of the tool, which

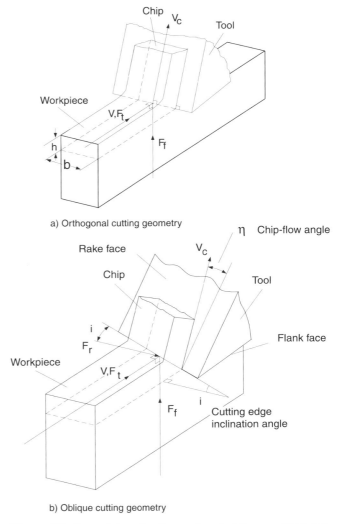

Figure 2.1: Geometries of orthogonal and oblique cutting processes.

is called the *sticking region*. The friction stress is approximately equal to the yield shear stress of the material at the sticking zone where the chip moves over a material stuck on the rake face of the tool. The chip stops sticking and starts sliding over the rake face with a constant *sliding friction* coefficient. The chip leaves the tool, losing contact with the rake face of the tool. The length of the contact zone depends on the cutting speed, tool geometry, and material properties. There are basically two types of assumptions in the analysis of the primary shear zone. Merchant [5] developed an orthogonal cutting model by assuming the shear zone to be a thin plane. Others, such as Lee and Shaffer [6] and Palmer and Oxley [7], based their analysis on a thick shear deformation

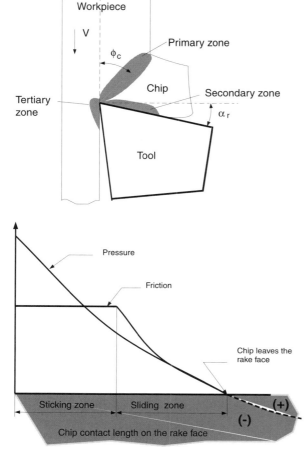

Figure 2.2: Deformation zones and distribution of load on the rake face. The tool experiences compression stress under the chip contact zone (−) and tensile stresses (+) after the chip leaves the tool.

zone, proposing "shear angle prediction" models in accordance with the laws of plasticity. In this text, the primary shear deformation zone is assumed to be a thin zone for simplification.

The deformation geometry and the cutting forces are shown on the cross-sectional view of the orthogonal cutting process (see Fig. 2.3). It is assumed that the cutting edge is sharp without a chamfer or radius and that the deformation takes place at the infinitely thin shear plane. The shear angle ϕ_c is defined as the angle between the direction of the cutting speed (V) and the shear plane. It is further assumed that the shear stress (τ_s) and the normal stress (σ_s) on the shear plane are constant; the resultant force (F) on the chip, applied at the shear plane, is in equilibrium to the force (F) applied to the tool over the chip–tool contact zone on the rake face; an average constant friction is assumed over the chip–rake face contact zone. From the force equilibrium, the resultant force (F) is formed from the feed (F_f) and tangential (F_t) cutting forces:

$$F = \sqrt{F_t^2 + F_f^2}. \qquad (2.1)$$

The feed force (or thrust force) is in the direction of uncut chip thickness and the tangential cutting force (or power force) is in the direction of cutting velocity. The cutting forces acting on the tool will have equal amplitude but opposite directions with respect to the forces acting on the chip. The mechanics of orthogonal cutting for two deformation zones are shown as follows:

Primary Shear Zone. The shear force (F_s) acting on the shear plane is derived from the geometry:

$$F_s = F \cos(\phi_c + \beta_a - \alpha_r), \qquad (2.2)$$

where β_a is average friction angle between the tool's rake face and the moving chip, and α_r is the rake angle of the tool. The shear force can also be expressed

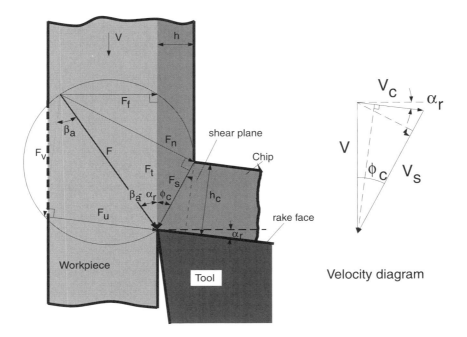

Cutting force diagram

Velocity diagram

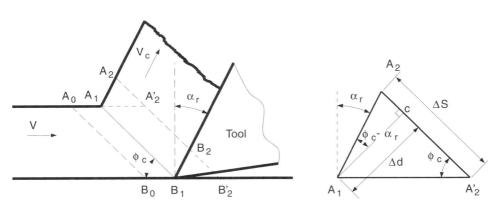

Shear deformation and strain diagrams

Figure 2.3: Mechanics of orthogonal cutting.

as a function of the feed and tangential cutting forces:

$$F_{\rm s} = F_{\rm t} \cos \phi_{\rm c} - F_{\rm f} \sin \phi_{\rm c}. \tag{2.3}$$

Similarly, the normal force acting on the shear plane is found to be

$$F_{\rm n} = F \sin(\phi_{\rm c} + \beta_{\rm a} - \alpha_{\rm r}) \tag{2.4}$$

or

$$F_n = F_t \sin \phi_c + F_f \cos \phi_c. \tag{2.5}$$

With the assumption of uniform stress distribution on the shear plane, the shear stress (τ_s) is found to be

$$\tau_s = \frac{F_s}{A_s}, \tag{2.6}$$

where the shear plane area (A_s) is

$$A_s = b\frac{h}{\sin \phi_c} \tag{2.7}$$

and b is the width of cut (or depth of cut in turning), h is the uncut chip thickness, and (ϕ_c) is the shear angle between the direction of cutting speed (V) and the shear plane. The normal stress on the shear plane (σ_s) is

$$\sigma_s = \frac{F_n}{A_s}. \tag{2.8}$$

The cutting velocity (V) is resolved into two components (see the velocity diagram shown in Fig. 2.3). The material is sheared away from the workpiece with the shear velocity (V_s). From the velocity hodograph shown, we have

$$V_s = V\frac{\cos \alpha_r}{\cos(\phi_c - \alpha_r)}. \tag{2.9}$$

The shear power spent in the shear plane is

$$P_s = F_s \cdot V_s, \tag{2.10}$$

which is converted into heat. The corresponding temperature rise on the shear plane (T_s) is

$$P_s = m_c c_s (T_s - T_r), \tag{2.11}$$

where m_c is the metal removal rate [kg/sec], c_s is the specific coefficient of heat for the workpiece material [Nm/kg°C], and T_r is the shop temperature. Metal removal rate is found from the cutting process conditions,

$$\left.\begin{array}{l} m_c = Q_c \rho, \\ Q_c = bhV \quad [\text{m}^3/\text{s}], \end{array}\right\} \tag{2.12}$$

where $\rho[\text{kg/m}^3]$ is the specific density of the workpiece material. The shear plane temperature (T_s) can be calculated from Eqs. (2.9) to (2.12):

$$T_s = T_r + \frac{P_s}{m_c c_s}. \tag{2.13}$$

The formulation given above considers that the entire plastic deformation takes place only at the shear plane and that all the heat is also consumed at the shear plane. This assumption is shown to overestimate the temperature prediction proposed by Boothroyd [8], who considered that some of the plastic deformation takes place over a shear zone of finite thickness and that some of the heat is dissipated to the work material and the chip, away from the thin shear plane. Oxley [4] used the following modified temperature prediction:

$$T_s = T_r + \lambda_h(1 - \lambda_s)\frac{P_s}{m_c c_s} \tag{2.14}$$

where λ_h $(0 < \lambda_h \leq 1)$ is a factor that considers the plastic work done outside the thin shear zone, and λ_s is the proportion of the heat conducted into the work material. For a plain carbon steel, an average value for $\lambda_h \approx 0.7$ can be assumed [9]. The heat conducted into the work material is evaluated with the following experimentally evaluated empirical equation [4]:

$$\begin{aligned} \lambda_s &= 0.5 - 0.35\log(R_T \tan\phi_c), && \text{for } 0.04 \leq R_T \tan\phi_c \leq 10, \\ \lambda_s &= 0.3 - 0.15\log(R_T \tan\phi_c), && \text{for } R_T \tan\phi_c \geq 10, \end{aligned} \tag{2.15}$$

where ϕ_c is the shear angle and R_T is a nondimensional thermal number given by

$$R_T = \frac{\rho c_s V h}{c_t}, \tag{2.16}$$

where c_t is the thermal conductivity of the work material with units [W/(m°C)]. Note also that the heat transmitted to the work material can not be more than the total energy generated, and a negative influx of the heat into the shear plane is not possible $(0 \leq \lambda_s \leq 1)$.

The shear plane length L_c is found from the chip deformation geometry,

$$L_c = \frac{h}{\sin\phi_c} = \frac{h_c}{\cos(\phi_c - \alpha_r)}. \tag{2.17}$$

The chip compression ratio (r_c) is the ratio of the uncut chip thickness over the deformed (h_c) one,

$$r_c = \frac{h}{h_c}. \tag{2.18}$$

The shear angle is found from the geometry as a function of rake angle and the chip compression ratio,

$$\phi_c = \tan^{-1}\frac{r_c \cos\alpha_r}{1 - r_c \sin\alpha_r}. \tag{2.19}$$

The shear strains and strain rates in metal cutting are significantly higher than those found from standard tensile tests and metal forming operations. The geometry of a deformed chip is shown in Figure 2.3. Assume that an undeformed chip section $A_0B_0A_1B_1$ is moving with workpiece velocity V. The workpiece material is deformed plastically at the shear plane (B_1A_1), and the cut chip slides over the rake face with a chip velocity V_c. After Δt shearing time, the uncut metal strip $A_0B_0B_1A_1$ becomes a chip with a geometry of $A_1B_1B_2A_2$. Hence, the chip is shifted from expected position of $B_2'A_2'$ to the deformed position B_2A_2 due to shearing in the shear plane with a shear angle of ϕ_c. Due to plane strain deformation, $A_2'A_2 = B_2'B_2$. The shear strain (γ_s) is defined as the ratio of deformation ($\Delta s = A_2'A_2$) over the nominal distance between the deformed and undeformed planes ($\Delta d = A_1C$),

$$\gamma_s = \frac{\Delta s}{\Delta d} = \frac{\overline{A_2 A_2'}}{\overline{A_1 C}} = \frac{\overline{A_2'C}}{\overline{A_1 C}} + \frac{\overline{CA_2}}{\overline{A_1 C}} = \cot\phi_c + \tan(\phi_c - \alpha_r).$$

By rearranging, the shear strain can be expressed as

$$\gamma_s = \frac{\cos\alpha_r}{\sin\phi_c \cos(\phi_c - \alpha_r)}. \tag{2.20}$$

The shear strain rate is

$$\gamma_s' = \frac{\gamma_s}{\Delta t}.$$

Assuming that the shear zone increment is Δs and that the thickness of shear deformation zone is Δd, the shear strain and shear velocity can be defined as $\gamma_s = \Delta s / \Delta d$ and $V_s = \Delta s / \Delta t$, respectively. The shear strain rate is then defined as

$$\gamma_s' = \frac{V_s}{\Delta d} = \frac{V \cos\alpha_r}{\Delta d \cos(\phi_c - \alpha_r)}. \tag{2.21}$$

Since the shear zone thickness Δd is extremely small in cutting, Eq. 2.21 indicates the presence of very high shear strain rates. Especially when the shear zone is assumed to be a plane with zero thickness, the strain rate becomes infinite, which can not be true. However, the thin shear plane approximation is useful for the macromechanics analysis of metal cutting. For practical and approximate predictions, the thickness of the shear zone can be approximated as a fraction of the shear plane length (i.e., $\Delta d \approx 0.15$–$0.2\,L_c$). For more accurate analysis, the shear zone thickness must be evaluated by freezing the machining process with a quick stop test and measuring the zone thickness with a scanning electron microscope (SEM).

Secondary Shear Zone. There are two components of the cutting force acting on the rake face of the tool (Fig. 2.3); the normal force F_v,

$$F_v = F_t \cos\alpha_r - F_f \sin\alpha_r, \tag{2.22}$$

and the friction force F_u on the rake face,

$$F_u = F_t \sin\alpha_r + F_f \cos\alpha_r. \tag{2.23}$$

In the orthogonal cutting analysis shown here, it is assumed that the chip is sliding on the tool with an average and constant friction coefficient of μ_a. In reality, the chip sticks to the rake face for a short period and then slides over the rake face with a constant friction coefficient [10]. The average friction coefficient on the rake face is given as

$$\mu_a = \tan\beta_a = \frac{F_u}{F_v}. \tag{2.24}$$

The friction angle β_a can alternatively be found from the tangential and feed forces,

$$\tan(\beta_a - \alpha_r) = \frac{F_f}{F_t}. \quad \rightarrow \quad \beta_a = \alpha_r + \tan^{-1}\frac{F_f}{F_t}. \tag{2.25}$$

The deformed chip slides on the rake face of the tool with the velocity of

$$V_c = r_c V = \frac{\sin\phi_c}{\cos(\phi_c - \alpha_r)}V. \tag{2.26}$$

Friction power spent on the tool chip contact face is

$$P_u = F_u V_c. \tag{2.27}$$

The total power consumed in cutting is the sum of energy spent in the shear and friction zones:

$$P_t = P_s + P_u. \tag{2.28}$$

From the equilibrium of cutting forces and the velocities, the total power is also equal to the cutting power drawn from the spindle motor:

$$P_t = F_t V. \tag{2.29}$$

The friction power increases the temperature of tool and chip. As can be seen from Eq. (2.27) if the velocity is increased, the friction power and thus the temperature of the tool increase. Excessive heat will cause undesirable high temperature in the tool, which leads to softening of the tool material and its accelerated wear and breakage. However, the production engineer desires an increased cutting velocity to obtain a high metal removal rate (Eq. 2.12) for productivity gains. The manufacturing researchers' challenge has been to decrease the cutting force F_u and move the heat toward the chip with better tool geometry design and to develop heat-resistant tool materials that can preserve their hardness at elevated temperatures. Although the prediction of the temperature distribution at the tool–chip interface is rather complex, the following simplified analysis is still useful for metal cutting engineers.

The friction power consumed at the tool chip interface (Eq. 2.27) is converted into heat via

$$P_u = m_c c_s \Delta T_c, \tag{2.30}$$

where ΔT_c is the average temperature rise in the chip. Boothroyd [8] and Stephenson [11] assumed a constant sticking friction load with a constant rectangular plastic zone at the tool–chip interface. The experimental temperature measurement and assumed plastic deformation zone led to the following empirical temperature relationship [4]:

$$\log\left(\frac{\Delta T_m}{\Delta T_c}\right) = 0.06 - 0.195\delta\sqrt{\frac{R_T h_c}{l_t}} + 0.5\log\left(\frac{R_T h_c}{l_t}\right), \tag{2.31}$$

where ΔT_m is the maximum temperature rise of the chip at the rake face–chip interface, which has a total contact length of l_t. The nondimensional number δ is the ratio of the plastic layer thickness over the deformed chip thickness (h_c) on the tool rake face–chip interface. The average temperature rise (T_{int}) at the rake face–chip interface is given by

$$T_{int} = T_s + \lambda_{int}\Delta T_m, \tag{2.32}$$

where T_s is the average shear plane temperature and λ_{int} (i.e., ≈ 0.7) is an empirical correction factor that accounts for temperature variations along the chip–tool contact zone. For an accurate analysis, both the plastic layer thickness (δh_c) and l_t must be measured with a microscope that has a large magnification (such as a SEM). Our experiments indicated that the thickness of the plastic layer on the rake face is observed to be between 5 and 10% of the deformed chip thickness $(\delta/h_c \approx 0.05$–$0.1)$. The contact length can be estimated approximately by assuming that the resultant cutting force acts in the middle of the contact length and parallel to the stress-free chip boundary. From the geometry of orthogonal cutting (Fig. 2.3), the chip–rake face contact length can be approximately predicted as

$$l_t = \frac{h\sin(\phi_c + \beta_a - \alpha_r)}{\sin\phi_c\cos\beta_a}. \tag{2.33}$$

The prediction of temperature distribution at the tool–chip interface is very important in determining the maximum speed that gives the most optimal material removal rate without excessive tool wear. The binding materials within the cutting tools may be weakened or diffused to the moving chip material at their critical diffusion or melting temperature limits. The fundamental machinability study requires the identification of a maximum cutting speed value that corresponds to the critical temperature limit where the tool wears rapidly. By using the approximate solutions summarized above, one can select a cutting speed that would correspond to a tool–chip interface temperature (T_{int}) that lies just below the diffusion and melting limits of materials present in a specific cutting tool. The detailed and fundamental scientific and experimental treatment of the cutting process is covered in Oxley [4].

It is difficult to predict the shear angle and stress in the shear plane and the average friction coefficient on the rake face using the standard material properties obtained from tensile and friction tests. For an accurate and realistic modeling, such fundamental parameters are identified from orthogonal cutting tests, where the deformed chip thickness and feed and tangential cutting forces are measured using cutting tools with a range of rake angles. The influence of uncut chip thickness and cutting speed is also considered by conducting experiments over a wide range of feeds and cutting speeds.

The relationships shown in Table 2.1 are identified from statistical analysis of more than 180 orthogonal cutting tests conducted using tungsten carbide (WC) cutting tools and Ti_6Al_4V titanium alloy work material. A set of turning experiments resembling orthogonal cutting was conducted on titanium tubes (Ti_6Al_4V) with tools of different rake angles at different feeds and cutting speeds. The diameter of the tube was 100 mm and the cutting speed range was 2.6 to 47 m/min. Cutting forces in the tangential (F_t) and feed (F_f) directions were measured with a force dynamometer. Two sample orthogonal cutting test results are shown in Figure 2.4. Small steps in cutting conditions were used to increase the reliability of the measured forces. It should be noted that the measured forces may include both the forces due to shearing and a tertiary deformation process "ploughing" or "rubbing" at the flank of the cutting edge. Thus the

TABLE 2.1. Orthogonal Cutting Data Base for Titanium Alloy Ti_6Al_4V
$\tau_s = 613$ (MPa)
$\beta_a = 19.1 + 0.29\alpha_r$ (deg)
$r_c = C_o h^{C_1}$
$C_o = 1.755 - 0.028\alpha_r$
$C_1 = 0.331 - 0.0082\alpha_r$
$K_{te} = 24$ (N/mm)
$K_{fe} = 43$ (N/mm)

measured force components are expressed as a superposition of shearing and edge forces:

$$F_t = F_{tc} + F_{te},$$
$$F_f = F_{fc} + F_{fe}. \tag{2.34}$$

The tests have been repeated a number of times at different feeds and cutting speed to ensure the statistical reliability of measurements. The edge forces are obtained by extrapolating the measured forces to zero chip thickness. It can be seen that the edge forces do not vary significantly with cutting speeds for the particular titanium alloy used here. The average edge force coefficients K_{te} and K_{fe} represent the rubbing forces per unit width. The chip compression ratio (r_c), shear stress τ_s, shear angle ϕ_c, and friction angle β_a (Table 2.1) are calculated from the measured "cutting" component of the forces and the cutting ratio by applying the orthogonal cutting theory presented above.

2.3 MECHANISTIC MODELING OF CUTTING FORCES

Orthogonal cutting mechanics are not directly applicable to many practical cutting tools with corner radius, side cutting edge angle, and chip breaking grooves. It is more practical to carry out a few experiments to identify constant parameters of the tool geometry–workpiece material pair to model existing cutting tools. However, it must be noted that for a tool design and analysis of a particular metal cutting process, oblique (i.e., three-dimensional) cutting mechanics and plasticity analysis are still necessary, and this is covered in Section 2.5.

As an example of mechanistic modeling, let us take the case of orthogonal cutting. We can extend the idea of model building to other cutting processes that are not orthogonal. In the previous section, the shearing force is formulated as a function of measured feed and tangential cutting forces in the orthogonal cutting. The shear force can be expressed as a function of shear stress and shear angle (Eqs. 2.6 and 2.7):

$$F_s = \tau_s b \frac{h}{\sin \phi_c}. \tag{2.35}$$

From Eqs. (2.2) and (2.35), the resultant cutting force (F) can be expressed in

Figure 2.4: Cutting forces measured during orthogonal turning of Ti_6Al_4V tubes with tungsten carbide tools.

terms of shear stress, friction and shear angles, width of cut, and feed rate:

$$F = \frac{F_s}{\cos(\phi_c + \beta_a - \alpha_r)} = \tau_s bh \frac{1}{\sin \phi_c \cos(\phi_c + \beta_a - \alpha_r)}. \tag{2.36}$$

The tangential and feed forces can be expressed in terms of resultant force:

$$\left.\begin{aligned} F_t &= F \cos(\beta_a - \alpha_r), \\ F_f &= F \sin(\beta_a - \alpha_r). \end{aligned}\right\} \tag{2.37}$$

Substituting Eq. (2.36) into Eq. (2.37), we can find the measured main cutting forces as functions of tool geometry and the cutting conditions (i.e., uncut chip thickness (h) and width of cut (a)) and process- and material-dependent terms (τ_s, β_a, ϕ_c, α_r):

$$F_t = bh\left[\tau_s\frac{\cos(\beta_a - \alpha_r)}{\sin\phi_c\cos(\phi_c + \beta_a - \alpha_r)}\right]. \tag{2.38}$$

Similarly the feed force is

$$F_f = bh\left[\tau_s\frac{\sin(\beta_a - \alpha_r)}{\sin\phi_c\cos(\phi_c + \beta_a - \alpha_r)}\right]. \tag{2.39}$$

In metal cutting literature the cutting parameter called *specific cutting pressure* or *tangential cutting force coefficient* (K_t) is defined as

$$K_t\,[\text{N/mm}^2] = \tau_s\frac{\cos(\beta_a - \alpha_r)}{\sin\phi_c\cos(\phi_c + \beta_a - \alpha_r)} \tag{2.40}$$

and the feed force constant (K_f) as

$$K_f\,[\text{N/mm}^2] = \tau_s\frac{\sin(\beta_a - \alpha_r)}{\sin\phi_c\cos(\phi_c + \beta_a - \alpha_r)}. \tag{2.41}$$

It is also customary to use another convention for cutting constants, where the feed force is assumed to be proportional to the tangential force with a ratio of

$$K_f = \frac{F_f}{F_t} = \tan(\beta_a - \alpha_r), \tag{2.42}$$

where K_f is dimensionless in this form. As can be seen from the definition (Eq. 2.40), the specific cutting pressure is a function of the yield shear stress of the workpiece (τ_s) material during cutting, the shear angle (ϕ_c), tool geometry (i.e., rake angle α_r), and the friction between the tool and the chip (β_a). In Eq. (2.40), only the tool geometry is known beforehand. The friction angle depends on the lubrication used, the tool–chip contact area, and the tool and workpiece materials. An accurate, analytical shear angle prediction remains the subject of continuing research. Previous research results are still insufficient to be used in predicting the shear angles accurately. The shear stress in the shear plane is also still in question with the present knowledge of the cutting process. If the shear plane is assumed to be a thick zone, which is more realistic than having a thin shear plane, there will be a work hardening, and the shear stress will be larger than the workpiece material's original yield shear stress measured from pure torsion or tensile tests. The temperature variation in the shear and the friction zones will also affect the hardness of the workpiece material; therefore the shear stress in the primary deformation zone will vary. The shear yield stress varies as a function of chip thickness as well due to varying strain hardening of the material being machined. Hence, it is customary to define the cutting forces mechanistically as a function of cutting conditions (i.e., b and h) and the cutting constants (K_{tc}) and (K_{fc}):

$$\left.\begin{array}{l} F_t = K_{tc}bh + K_{te}b, \\ F_f = K_{fc}bh + K_{fe}b. \end{array}\right\} \tag{2.43}$$

The cutting constants (K_{tc}, K_{fc}), and the edge coefficients that do not contribute to the shearing (K_{te}, K_{fe}), are directly calibrated from metal cutting experiments for a tool–workpiece pair. Note that the edge coefficients change as the cutting tool wears or experiences chipping. It should be also noted that in order to take the influence of the chip thickness on the friction and shear angles, and the yield shear stress, the specific cutting pressure (K_t) and ratio (K_f) are sometimes expressed as nonlinear functions of uncut chip thickness:

$$\left.\begin{array}{l} K_t = K_T h^{-p}, \\ K_f = K_F h^{-q}, \end{array}\right\} \tag{2.44}$$

where p and q are cutting force constants determined from the cutting experiments at different feed rates. Equation (2.44) represents basic nonlinearity in the cutting force expressions. This form is used when the edge forces are neglected in the mechanistic models. It must be noted that some work materials exhibit different yield stress and friction coefficient at different speeds, which lead to the speed dependency of cutting constants. The mechanistic cutting constant equation (2.44) can be extended to include cutting speed as a variable.

   Example. The cutting conditions for turning an AISI-1045 steel workpiece are set as follows: depth of cut $b = 2.54$ mm; feed rate $c = 0.2$ mm/rev; spindle speed $n = 350$ rev/min; workpiece diameter $= 100.$ mm; tool's rake angle $\alpha_r = +5°$. Specific mass of the steel $\rho = 7{,}800$ kg/m^3; specific heat coefficient of steel $c_s = 470$ Nm/kg°C; thermal conductivity $c_t = 28.74$ [W/m°C]. The following measurements are observed from the experiment: deformed chip thickness $h_c = 0.44$ mm, feed force $F_f = 600$ N, tangential force $F_t = 1{,}200$ N. Assuming that the turning is an orthogonal metal cutting process, the following values are evaluated:

Resultant cutting force	$F = \sqrt{F_t^2 + F_f^2} = 1342.$ N
Chip ratio	$r_c = \frac{h}{h_c} = 0.4545$
Shear angle	$\phi_c = \tan^{-1}\frac{r_c \cos\alpha_r}{1 - r_c \sin\alpha_r} = 25°$
Friction angle	$\beta_a = \alpha_r + \tan^{-1}\frac{F_f}{F_t} = 31.6°$
Friction coefficient	$\mu_a = \tan\beta_a = 0.6144$
Shearing force	$F_s = F\cos(\phi_c + \beta_a - \alpha_r) = 833.5$ N
Shear plane area	$A_s = b\frac{h}{\sin\phi_c} = 1.2$ mm^2
Shearing stress	$\tau_s = \frac{F_s}{A_s} = 693.4$ MPa
Normal force on the shear plane	$F_n = F\sin(\phi_c + \beta_a - \alpha_r) = 1051.7$ N
Normal stress on the shear plane	$\sigma_s = \frac{F_n}{A_s} = 876.43$ MPa

Cutting speed	$V = \pi D n = 110$ m/min
Shearing velocity	$V_s = V \frac{\cos \alpha_r}{\cos(\phi_c - \alpha_r)} = 116.6$ m/min $= 1.9436$ m/s
Shearing power	$P_s = F_s V_s = 1{,}620$ W
Metal removal rate	$m_c = Q_c \rho = bhV\rho = 7.2644 \cdot 10^{-3}$ kg/s
Nondimensional thermal number	$R_T = \frac{\rho c_s V h}{c_t} = 45.78,\ \ R_T \tan \phi_c = 21.34 > 10$
Scale of heat conducted into work	$\lambda_s = 0.3 - 0.15 \log(R_T \tan \phi_c) = 0.1$
Shear plane temperature	$T_s = T_r + \lambda_h (1 - \lambda_s)\frac{P_s}{m_c c_s} = 20 + 299 = 319°$C,
	$(\lambda_h \approx 0.7)$
Friction force	$F_u = F \sin \beta_a = 703.2$ N
Normal force	$F_v = F \cos \beta_a = 1143$ N
Chip velocity	$V_c = r_c V = 50$ m/min $= 0.8333$ m/s
Friction power	$P_u = F_u V_c = 586$ W
Chip contact length	$l_t = \frac{h \sin(\phi_c + \beta_a - \alpha_r)}{\sin \phi_c \cos \beta_a} = 0.435$ mm
Total cutting power drawn	$P_t = P_u + P_s = 2{,}200$ W
Specific cutting pressure	$K_t = \frac{F_t}{bh} = 2{,}362.$ N/mm^2
Cutting force ratio	$K_f = \frac{F_f}{F_t} = 0.5$

2.4 THEORETICAL PREDICTION OF SHEAR ANGLE

The evaluation of shear angle, shear stress, and average friction coefficient from orthogonal metal cutting tests was summarized in the previous sections. There have been many attempts in predicting the shear angle theoretically, without relying on metal cutting experiments. Some of the most fundamental models, which assume a perfect rigid plastic workpiece material without any strain hardening, are briefly presented in this section. These models assume that the shear plane is thin; that the shear stress in the shear plane is equivalent to the yield shear stress of the material; and that the average friction is found from friction tests between the tool and workpiece materials, leaving only the shear angle as unknown. There have been two fundamental approaches to predict the shear angle as follows.

Maximum Shear Stress Principle

Krystof [12] proposed a shear angle relation based on the maximum shear stress principle (i.e., shear occurs in the direction of maximum shear stress). The resultant force makes an angle $(\phi_c + \beta_a - \alpha_r)$ with the shear plane (see Fig. 2.3), and the angle between the maximum shear stress and the principal stress (i.e., the resultant force) must be $\pi/4$. Therefore, the following shear angle relation is obtained:

$$\phi_c = \frac{\pi}{4} - (\beta_a - \alpha_r). \tag{2.45}$$

Later, Lee and Shaffer [6] derived the same shear angle relationship from a slip-line field model.

Minimum Energy Principle

Merchant [13] proposed applying the minimum energy principle in predicting the shear angle. By taking the partial derivative of the cutting power (Eqs. 2.29 and 2.38), one gets

$$\frac{dP_t}{d\phi_c} = \frac{d(VF_t)}{d\phi_c} = \frac{-V\tau_s bh \cos(\beta_a - \alpha_r) \cos(2\phi_c + \beta_a - \alpha_r)}{\sin^2 \phi_c \cos^2(\phi_c + \beta_a - \alpha_r)} = 0,$$

or, with $\cos(2\phi_c + \beta_a - \alpha_r) = 0$,

$$\phi_c = \frac{\pi}{4} - \frac{\beta_a - \alpha_r}{2}, \tag{2.46}$$

which predicts a larger shear angle than the maximum shear stress principle. Although the equations above proposed by Krystof, Merchant, and Lee-Shaffer do not yield qualitatively accurate shear angle predictions owing to oversimplified assumptions, they provided an important relationship among the shear angle (ϕ_c), the rake angle (α_r), which is the most fundamental for tool design, and the friction coefficient between the workpiece and cutting tool materials ($\tan \beta_a$). The forces and power consumed in cutting decrease with increasing shear angle. The expression indicates that the friction coefficient between the tool and the chip must be decreased by using lubricants or materials with a smaller friction coefficient, and the rake angle of the cutting tool must be increased as much as possible provided the weakened cutting edge is able to withstand the pressure and friction load exerted by the chip at the rake face contact zone.

2.5 MECHANICS OF OBLIQUE CUTTING

The geometry of oblique cutting is shown in Figure 2.5. The cutting velocity (V) is perpendicular to the cutting edge in orthogonal cutting, whereas in oblique cutting, it is inclined at an acute angle i to the plane normal to the cutting edge.

2.5.1 Oblique Cutting Geometry

The difference in the geometry of two cutting mechanisms may be better explained by revisiting the orthogonal cutting geometry shown in Figure 2.1. A plane normal to the cutting edge and parallel to the cutting velocity V is defined as the *normal plane* or P_n. Since the shear deformation is plane strain without side spreading, the shearing and chip motion are identical on all the normal planes parallel to the cutting speed V and perpendicular to the cutting edge. Hence, the velocities of cutting (V), shear (V_s), and chip (V_c) are all perpendicular to the cutting edge, and they lie in the velocity plane (P_v) parallel to or coincident with the normal plane (P_n). The resultant cutting force F, along with the other forces acting on the shear and chip–rake face contact zone, also lies on the same plane P_n in orthogonal cutting. There is no cutting force in the third direction (i.e., perpendicular to the normal plane). However, the cutting velocity has an *oblique* or inclination angle i in oblique cutting operations, and thus the directions of shear, friction, chip flow, and resultant cutting force vectors have components in all three Cartesian coordinates (x, y, z) (see Fig. 2.5). In Figure 2.5, the x axis is perpendicular to the cutting edge but lies on the cut surface; y is aligned with the cutting edge; and z is perpendicular to the xy plane. The forces exist in all three directions in oblique cutting. The important planes in oblique cutting are the shear plane, the rake face, the cut surface xy, the *normal plane xz* or P_n, and the *velocity plane* P_v. Most analyses assume that the mechanics of oblique cutting in the normal plane are equivalent to that of orthogonal cutting; hence all velocity and force vectors are projected on the normal plane. In Figure 2.6, the angle between the shear and xy planes is called the *normal shear angle* ϕ_n. The shear velocity lies on the shear plane but makes an *oblique shear angle* ϕ_i with the vector normal to the cutting edge on the normal plane. The sheared chip moves over the rake face plane with a *chip flow angle* η measured from a vector

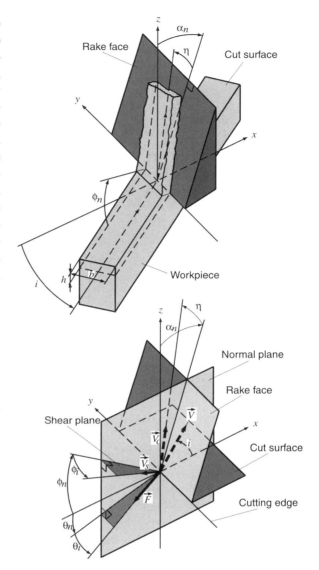

Figure 2.5: The geometry of oblique cutting process.

on the rake face but normal to the cutting edge. Note that this normal vector also lies on the normal plane P_n. The friction force between the chip and the rake face is collinear with the direction of chip flow. The angle between the z axis and normal vector on the rake face is defined as the *normal rake angle* α_n. The friction force on the rake face ($\vec{F}_u$) and normal force to the rake ($\vec{F}_v$) form the resultant cutting force $\vec{F}$ with a friction angle of β_a (see Fig. 2.6). The resultant force vector ($\vec{F}$) has an acute projection angle of θ_i with the normal plane P_n, which in turn has an in-plane angle of $\theta_n + \alpha_n$ with the normal force F_v. Here, θ_n is the angle between the x axis and the projection of $\vec{F}$ on P_n. The following geometric relations can be derived from Figure 2.6:

$$F_{\mathrm{u}} = F\sin\beta_{\mathrm{a}} = F\frac{\sin\theta_i}{\sin\eta} \;\rightarrow\; \sin\theta_i = \sin\beta_{\mathrm{a}}\sin\eta, \tag{2.47}$$

$$F_{\mathrm{u}} = F_{\mathrm{v}}\tan\beta_{\mathrm{a}} = F_{\mathrm{v}}\frac{\tan(\theta_{\mathrm{n}}+\alpha_{\mathrm{n}})}{\cos\eta} \;\rightarrow\; \tan(\theta_{\mathrm{n}}+\alpha_{\mathrm{n}}) = \tan\beta_{\mathrm{a}}\cos\eta. \tag{2.48}$$

The velocities of chip ($\vec{V}_{\mathrm{c}}$), shear ($\vec{V}_{\mathrm{s}}$), and cutting ($\vec{V}$) form the velocity plane P_{v} as shown in Figure 2.6. Each velocity vector can be defined by its Cartesian components:

$$\begin{aligned}
\vec{V} &= (V\cos i, & V\sin i, & \quad 0),\\
\vec{V}_{\mathrm{c}} &= (V_{\mathrm{c}}\cos\eta\sin\alpha_{\mathrm{n}}, & V_{\mathrm{c}}\sin\eta, & \quad V_{\mathrm{c}}\cos\eta\cos\alpha_{\mathrm{n}}),\\
\vec{V}_{\mathrm{s}} &= (-V_{\mathrm{s}}\cos\phi_i\cos\phi_{\mathrm{n}}, & -V_{\mathrm{s}}\sin\phi_i, & \quad V_{\mathrm{s}}\cos\phi_i\sin\phi_{\mathrm{n}}),
\end{aligned}$$

By eliminating V, V_{c}, and V_{s} from the velocity relation

$$\vec{V}_{\mathrm{s}} = \vec{V}_{\mathrm{c}} - \vec{V},$$

the following geometric relation between the shear and the chip flow directions can be obtained [14]:

$$\tan\eta = \frac{\tan i\,\cos(\phi_{\mathrm{n}}-\alpha_{\mathrm{n}}) - \cos\alpha_{\mathrm{n}}\tan\phi_i}{\sin\phi_{\mathrm{n}}}. \tag{2.49}$$

The above relationships define the geometry of the oblique cutting process.

2.5.2   Solution of Oblique Cutting Parameters

There are five unknown oblique cutting parameters, which define the directions of resultant force ($\theta_{\mathrm{n}}, \theta_i$), shear velocity ($\phi_{\mathrm{n}}, \phi_i$), and chip flow ($\eta$). In addition to the three equations (2.47–2.49) obtained from the oblique cutting geometry, two additional expressions are required to solve for the five unknown angles. There have been numerous proposed solutions based on the empirical chip flow direction [2, 15] and other empirical assumptions [16–18]. A theoretical shear angle prediction approach has been proposed by Shamoto and Altintas [19]; it parallels the maximum shear stress [6, 12] and the minimum energy [13] principles used in two-dimensional orthogonal cutting mechanics. The focus here is on the prediction of shear direction based on the laws of mechanics, not on the combined empirical predictions and geometric relations.

Maximum Shear Stress Principle. Krystof [12] applied the maximum shear stress criterion to predict the direction of shear angle in orthogonal cutting (i.e., $\phi_{\mathrm{n}} = \pi/4 - \beta + \alpha_{\mathrm{n}}$). Later, Lee and Shaffer [6] achieved the same orthogonal shear angle relationship using the slip-line field solution. Both assume that the shear occurs in the direction of maximum shear stress (where the angle between the shear velocity and resultant force directions is 45°; see Figure 2.6). The same principle can be applied to oblique cutting, that is, *the resultant force*

$(\vec{F})$ *makes a* 45° *acute angle with the direction of shear*:

$$F_s = F(\cos\theta_i \cos(\theta_n+\phi_n)\cos\phi_i$$
$$+ \sin\theta_i \sin\phi_i) = F\cos(45°).$$

Furthermore, the same principle dictates that the projection of $\vec{F}$ to the shear plane coincides with the shear direction, that is, *the component of the resultant force in the direction normal to the shear on the shear plane must be zero*:

$$F(\cos\theta_i \cos(\theta_n+\phi_n)\sin\phi_i$$
$$- \sin\theta_i \cos\phi_i) = 0;$$

otherwise the shear stress in the shear direction is not the maximum on the shear plane. The two expressions provide the necessary relationships between the shear and resultant force directions:

$$\sin\phi_i = \sqrt{2}\sin\theta_i, \tag{2.50}$$

$$\cos(\phi_n + \theta_n) = \frac{\tan\theta_i}{\tan\phi_i}. \tag{2.51}$$

By solving the five equations (2.47) to (2.51), the five unknown angles (ϕ_n, ϕ_i, θ_n, θ_i, η) that describe the mechanics of oblique cutting, can be obtained. However, direct analytical solution of the equations are rather difficult; hence they are solved by employing an iterative numerical method. The numerical solution is obtained according to the block diagram shown in Figure 2.7. Note that the angles of friction (β_a), rake (α_n), and inclination (i) are known from the geom-

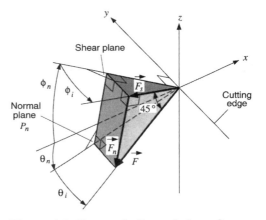

Figure 2.6: Force, velocity, and shear diagrams in oblique cutting.

etry and material tests and are considered as inputs to the system. The iterative solution is started by assuming an initial value for the chip flow angle (i.e., $\eta = i$) as proposed by Stabler [20]. The direction (θ_n, θ_i) of the resultant force vector $\vec{F}$ is obtained from Eqs. (2.47) and (2.48). Similarly, the shear direction angles (ϕ_n, ϕ_i) can be evaluated from Eqs. (2.50) and (2.51), followed by evaluating the new chip flow angle η_e from the velocity equation (2.49). The true chip flow angle is searched iteratively using the following interpolation algorithm:

$$\eta(k) = \nu\,\eta(k-1) + (1-\nu)\,\eta_e,$$

Figure 2.7: Solution procedure.

where k is the iteration counter, and the interpolation ratio ν is selected within the range of $0 < \nu < 1$. The ratio ν is dynamically updated for the rapid convergence, that is, decreased if $\eta(k)$ oscillates and increased if its value moves in the same direction. The iteration is continued until the chip flow angle converges within $10^{-12}\%$. When the three-dimensional oblique cutting model introduced above is applied to two-dimensional orthogonal cutting, it yields the same shear angle expression proposed by Krystof [12] and by Lee and Shaffer [6] who used the maximum shear stress principle on orthogonal cutting (i.e., $i = \theta_i = \phi_i = 0 \rightarrow \phi_n = \pi/4 - (\beta_a - \alpha_n)$). Note that with identity (2.50), Eq. (2.51) reduces to $\cos(\phi_n + \theta_n) = 1/\sqrt{2}$ in orthogonal cutting.

Minimum Energy Principle. Merchant [13] proposed a shear angle prediction theory by applying the minimum energy principle to orthogonal cutting. The same principle is extended to oblique cutting here. From the geometry, the shear force is expressed as a projection of $\vec{F}$ in the direction of shear (Fig. 2.6):

$$F_s = F[\cos(\theta_n + \phi_n)\cos\theta_i \cos\phi_i + \sin\theta_i \sin\phi_i]$$

or as a product of shear stress and shear plane area (Fig. 2.5):

$$F_s = \tau_s A_s = \tau \left(\frac{b}{\cos i}\right)\left(\frac{h}{\sin\phi_n}\right),$$

where A_s, b, and h are the shear area, the width of cut, and the depth of cut (uncut chip thickness), respectively. By equating the two shear force expressions, the resultant force is derived:

$$F = \frac{\tau_s bh}{[\cos(\theta_n + \phi_n)\cos\theta_i \cos\phi_i + \sin\theta_i \sin\phi_i]\cos i \sin\phi_n}. \tag{2.52}$$

The cutting power (P_t) in oblique cutting can be expressed as a function of F (see Fig. 2.6):

$$P_t = F_t V = F(\cos\theta_i \cos\theta_n \cos i + \sin\theta_i \sin i)V,$$

and the nondimensional power (P_t') is obtained by substituting Eq. (2.52) for F.

This gives

$$P'_t = \frac{P_t}{V\tau_s bh} = \frac{\cos\theta_n + \tan\theta_i \tan i}{[\cos(\theta_n + \phi_n)\cos\phi_i + \tan\theta_i \sin\phi_i]\sin\phi_n}, \tag{2.53}$$

where the term $V\tau_s bh$ is constant. The minimum energy principle requires that the cutting power must be minimum for a unique shear angle solution. Since the direction of shear is characterized by the shear angles ϕ_n and ϕ_i, we have

$$\begin{aligned} \partial P'_t / \partial\phi_n &= 0, \\ \partial P'_t / \partial\phi_i &= 0, \end{aligned} \tag{2.54}$$

which provides two more equations in addition to the three geometric relations given by Eqs. (2.47), (2.48), and (2.49). Hence, the five unknown angles $(\phi_n, \phi_i, \theta_n, \theta_i, \eta)$, which describe the mechanics of oblique cutting, can be obtained. However, because finding an analytical solution of five equations is rather difficult, the solution is obtained using a numerical iteration technique. The algorithm starts with an initial value of chip flow angle $\eta = i$ (Stabler [20]), followed by the evaluation of the remaining angles from Eqs. (2.47), (2.48), (2.50), and (2.51). After calculating the initial values of $\theta_n, \theta_i, \phi_n$, and ϕ_i, the cutting power (P'_t) is obtained from Eq. (2.53). By changing the shear angles slightly (i.e., $\phi_n + \Delta\phi_n$ and $\phi_i + \Delta\phi_i$), the steepest descent direction $(\Delta P'_t / \Delta\phi_n, \Delta P'_t / \Delta\phi_i)$ is evaluated. The shear angles are changed by a step ζ in the steepest descent direction so that the cutting energy approaches the minimum value:

$$\begin{Bmatrix} \phi_n(k) \\ \phi_i(k) \end{Bmatrix} = \begin{Bmatrix} \phi_n(k-1) \\ \phi_i(k-1) \end{Bmatrix} - \zeta \begin{Bmatrix} \Delta P'_t / \Delta\phi_n \\ \Delta P'_t / \Delta\phi_i \end{Bmatrix}.$$

The numerical iteration is continued in accordance with Figure 2.7 until the nondimensional cutting power (P'_t) converges to a minimum value. When the three-dimensional oblique cutting model introduced above is applied to two-dimensional orthogonal cutting, it yields the same shear angle expression proposed by Merchant [13] (i.e., $i = \theta_i = \phi_i = 0 \rightarrow \phi_n = \pi/4 - (\beta_a - \alpha_n)/2$).

Empirical Approach. There are a number of empirical models and we introduce the one proposed by Armarego [21] here. In his model, two assumptions are employed on the shear direction and chip length ratio: (1) the shear velocity is collinear with shear force and (2) the chip length ratio in oblique cutting is the same as that in orthogonal cutting. The former assumption made by Stabler [20] is considered as one of the maximum shear stress criteria (Eq. 2.51), and by combining the three geometric equations derived in the previous section, it yields

$$\tan(\phi_n + \beta_n) = \frac{\cos\alpha_n \tan i}{\tan\eta - \sin\alpha_n \tan i}, \tag{2.55}$$

where $\beta_n = \theta_n + \alpha_n$. Thus it is given by the following equation (see Eq. (2.48)):

$$\tan\beta_n = \tan\beta_a \cos\eta. \tag{2.56}$$

Armarego [21] made the latter assumption from his experiments, so that the normal shear angle ϕ_n can be obtained from the following equation of chip geometry:

$$\tan(\phi_n) = \frac{r_c(\cos \eta / \cos i) \cos \alpha_n}{1 - r_c(\cos \eta / \cos i) \sin \alpha_n}. \tag{2.57}$$

By solving these three equations numerically, the three unknown angles η, ϕ_n, and β_n are derived, or numerical iteration can be avoided if Stabler's [20] empirical chip flow rule (i.e., $\eta = i$) is applied to Eq. (2.57).

2.5.3 Prediction of Cutting Forces

The cutting force components are derived as projections of the resultant cutting force F. They are expressed as a function of shear yield stress τ_s, resultant force direction (θ_n, θ_i), oblique angle i, and obique shear angles (ϕ_i, ϕ_n) as given in Eq. (2.52) for F. The force components in the directions of cutting speed (F_t), the thrust (F_f), and the normal (F_r) are given by the following equations (see Figs. 2.5 and 2.6):

$$\left. \begin{aligned} F_t &= F(\cos \theta_i \cos \theta_n \cos i + \sin \theta_i \sin i) \\ &= \frac{\tau_s bh(\cos \theta_n + \tan \theta_i \tan i)}{[\cos(\theta_n + \phi_n) \cos \phi_i + \tan \theta_i \sin \phi_i] \sin \phi_n}, \\ F_f &= F \cos \theta_i \sin \theta_n \\ &= \frac{\tau_s bh \sin \theta_n}{[\cos(\theta_n + \phi_n) \cos \phi_i + \tan \theta_i \sin \phi_i] \cos i \sin \phi_n}, \\ F_r &= F(\sin \theta_i \cos i - \cos \theta_i \cos \theta_n \sin i) \\ &= \frac{\tau_s bh(\tan \theta_i - \cos \theta_n \tan i)}{[\cos(\theta_n + \phi_n) \cos \phi_i + \tan \theta_i \sin \phi_i] \sin \phi_n}. \end{aligned} \right\} \tag{2.58}$$

It is convenient to express the cutting forces in the following form:

$$\left. \begin{aligned} F_t &= K_{tc}bh + K_{te}b, \\ F_f &= K_{fc}bh + K_{fe}b, \\ F_r &= K_{rc}bh + K_{re}b, \end{aligned} \right\} \tag{2.59}$$

where the cutting constants contributed by the shear action correspond to

$$\left. \begin{aligned} K_{tc} &= \frac{\tau_s(\cos \theta_n + \tan \theta_i \tan i)}{[\cos(\theta_n + \phi_n) \cos \phi_i + \tan \theta_i \sin \phi_i] \sin \phi_n}, \\ K_{fc} &= \frac{\tau_s \sin \theta_n}{[\cos(\theta_n + \phi_n) \cos \phi_i + \tan \theta_i \sin \phi_i] \cos i \sin \phi_n}, \\ K_{rc} &= \frac{\tau_s(\tan \theta_i - \cos \theta_n \tan i)}{[\cos(\theta_n + \phi_n) \cos \phi_i + \tan \theta_i \sin \phi_i] \sin \phi_n}. \end{aligned} \right\} \tag{2.60}$$

If Armarego's classical oblique model is employed, the force expressions can be transformed into the following by using the geometric relations explained above:

$$
\left.
\begin{aligned}
F_t &= bh \cdot \left[\frac{\tau_s}{\sin\phi_n} \frac{\cos(\beta_n - \alpha_n) + \tan i \tan\eta \sin\beta_n}{\sqrt{\cos^2(\phi_n + \beta_n - \alpha_n) + \tan^2\eta \sin^2\beta_n}} \right], \\[2ex]
F_f &= bh \cdot \left[\frac{\tau_s}{\sin\phi_n \cos i} \frac{\sin(\beta_n - \alpha_n)}{\sqrt{\cos^2(\phi_n + \beta_n - \alpha_n) + \tan^2\eta \sin^2\beta_n}} \right], \\[2ex]
F_r &= bh \cdot \left[\frac{\tau_s}{\sin\phi_n} \frac{\cos(\beta_n - \alpha_n)\tan i - \tan\eta \sin\beta_n}{\sqrt{\cos^2(\phi_n + \beta_n - \alpha_n) + \tan^2\eta \sin^2\beta_n}} \right].
\end{aligned}
\right\}
\tag{2.61}
$$

Hence, the corresponding cutting constants are

$$
\left.
\begin{aligned}
K_{tc} &= \frac{\tau_s}{\sin\phi_n} \frac{\cos(\beta_n - \alpha_n) + \tan i \tan\eta \sin\beta_n}{\sqrt{\cos^2(\phi_n + \beta_n - \alpha_n) + \tan^2\eta \sin^2\beta_n}}, \\[2ex]
K_{fc} &= \frac{\tau_s}{\sin\phi_n \cos i} \frac{\sin(\beta_n - \alpha_n)}{\sqrt{\cos^2(\phi_n + \beta_n - \alpha_n) + \tan^2\eta \sin^2\beta_n}}, \\[2ex]
K_{rc} &= \frac{\tau_s}{\sin\phi_n} \frac{\cos(\beta_n - \alpha_n)\tan i - \tan\eta \sin\beta_n}{\sqrt{\cos^2(\phi_n + \beta_n - \alpha_n) + \tan^2\eta \sin^2\beta_n}}.
\end{aligned}
\right\}
\tag{2.62}
$$

The following practical approach can be used in predicting the oblique cutting forces from the orthogonal cutting database [18]:

- Evaluate the shear angle (ϕ_c), average friction angle (β_a), and shear yield stress (τ_s) from orthogonal cutting tests (i.e., as given in Table 2.1).
- The following assumptions are made: The orthogonal shear angle is equal to the normal shear angle in oblique cutting ($\phi_c \equiv \phi_n$); the normal rake angle is equal to the rake angle in orthogonal cutting ($\alpha_r \equiv \alpha_n$); the chip flow angle is equal to the oblique angle ($\eta \equiv i$) by adopting Stabler's chip flow rule [20]; the friction coefficient (β_a) and shear stress (τ_s) are the same in both orthogonal and oblique cutting operations for a given speed, chip load, and tool–work material pair.
- Predict the cutting forces using the oblique cutting constants given in Eq. (2.62).

Many practical oblique cutting operations, such as turning, drilling, and milling, can be evaluated using the oblique cutting mechanics procedure outlined above.

2.6 MECHANICS OF TURNING PROCESSES

A typical conventional lathe is shown in Figure 2.8. The rotational workpiece is held in a chuck, which is bolted to the spindle, and a single point cutting tool is attached to a tool post. A turning operation is used to machine cylindrical parts (see Fig. 2.9). When the workpiece is long and heavy, its two ends are

Figure 2.8: Conventional lathe.

Figure 2.9: A profile CNC turning operation with a carbide tool. Source: Mitsubishi Materials Corp.

held by the chuck and the center of the tail stock, respectively. The tool post is held on the top of a carriage, which has motions along the axis between the spindle and tail stock centers and is perpendicular to this axis. Conventional lathes have a single motor with constant speed. The speed is transmitted to spindle and feed drive gear boxes via belts. The speed is further reduced using gear combinations within the feed and spindle drive boxes, which have shift arms with speed labels. However, in CNC lathes the spindle and feed speeds are directly programmed within NC programs, since they have computer controlled stepless drives with one or zero gear reduction level. A Computer Numerical Control (CNC) turning center with a turret carrying multiple cutting tools is shown in Figure 2.10. The turret, which holds the tool, can be moved along or perpendicular to the spindle axis or both in CNC lathes. If the tool moves along the main axis, the tool reduces the diameter of the cylindrical workpiece. If it moves perpendicular to the main axis, the tool removes material from the flat face of the workpiece, which is called a *facing* operation. The combination of cylindrical turning and facing can be used for chamfering and parting operations. The lathes allow synchronized linear motion of the carriage feed and the rotational motion of the spindle via gear boxes. The synchronized motion is used for threading operations.

The diagram of a typical cylindrical turning process is shown in Figure 2.11. The cutting tool moves parallel to the spindle and removes a skin from the blank, hence reducing the diameter of the shaft. A standard turning tool geometry is shown in Figure 2.12. The important geometric parameters on the tool are tool nose radius, side rake, back rake, and side cutting edge angles. The chip lands and slides on the rake face of the tool. Side rake angle is the

Figure 2.10: A CNC turning center. Source: Hitachi Seiki Co., Ltd.

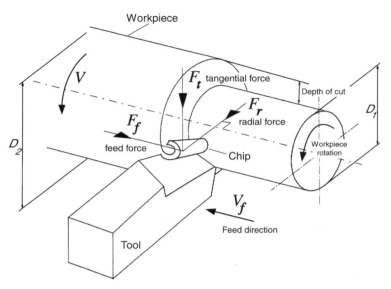

Figure 2.11: Geometry of turning process.

inclination of the rake face toward the cutting edge, whereas the back rake angle indicates the inclination toward the tip of the tool that is perpendicular to the surface of the workpiece in turning. In orthogonal cutting, there is zero back rake, and only side rake is considered. The tools are called positive, neutral (i.e., zero rake angle), or negative depending on the orientation of the rake angles. Positive rake angle produces higher shear angles; therefore it helps to reduce the cutting forces. It also leaves a better surface finish since it assists the chip to flow away from the workpiece. Negative tools produce higher force than the positive tools for the same cutting conditions, because the negative rake angles decrease the shear angle. However, in interrupted cutting, where the tool periodically enters into and exits from the workpiece, the negative tools provide a greater shock resistance than positive tools since the initial material contact with the tool occurs away from the weak cutting edge. Usually, cutting tool inserts made of carbide, ceramic, diamond, or Cubic Boron Nitride (CBN) are clamped on tool holders that have a similar geometry like the one shown in Figure 2.12. Only one side of the positive tools can be used, because of the clearance angle at the edges. The negative carbide inserts have zero clearance angles; therefore both sides of the insert can be used, which lowers tool costs. For example, eight edges of a negative but only four edges of a square positive insert can be used in machining. The tools have a small radius at their noses to minimize sharp feed marks on the finished surface. A large tool nose radius is not advisable because it makes the tool susceptible to self-excited vibrations, or chatter, in machining.

There are various definitions of tool angles. However, the following tool angle conversions based on back rake–side rake [2] are the most commonly accepted

Figure 2.12: Geometry of turning tool.

for analyzing the mechanics of oblique cutting:

$$\left.\begin{aligned}
\tan\alpha_{\mathrm{o}} &= \tan\alpha_{\mathrm{f}}\cos\psi_{\mathrm{r}} + \tan\alpha_{\mathrm{p}}\sin\psi_{\mathrm{r}}, \\
\tan i &= \tan\alpha_{\mathrm{p}}\cos\psi_{\mathrm{r}} + \tan\alpha_{\mathrm{f}}\sin\psi_{\mathrm{r}}, \\
\tan\alpha_{\mathrm{n}} &= \tan\alpha_{\mathrm{o}}\cos i,
\end{aligned}\right\} \tag{2.63}$$

where i, α_{o}, and α_{n} are the equivalent oblique, orthogonal, and normal rake angles, respectively. The equivalent oblique (i) and normal rake (α_{n}) angles must be evaluated for oblique cutting tools before using the cutting constants given in Eq. (2.62) for predicting the cuting forces. For simplicity, the chip

Figure 2.13: Mechanics of turning with bull-nosed inserts.

flow angle can be assumed to be equal to the oblique angle as suggested by Stabler [20].

Prediction of Cutting Forces in Turning. The cutting forces in turning can be predicted by transforming the orthogonal cutting parameters to an oblique turning geometry using the angle transformations given in Eq. (2.63) and the oblique cutting constants given in Eq. (2.62). The process can be explained by taking a turning tool with a corner radius r, side rake angle of α_f, and back rake angle of α_p as shown in Figure 2.13. The diameter of the workpiece is d, and the radial depth of cut is a, which is larger than the tool nose radius (r). The feed rate is c, which corresponds to the linear travel of the cutting tool in one revolution of the spindle. The chip has a uniform thickness in region I where the radial depth is less than the radius (r). However, the chip has a varying thickness in region II owing to the corner radius of the tool. The cutting forces can be predicted by applying the classical oblique cutting transformation proposed by Armarego ([22,23]) as explained in Section 2.5. The transformation is applied separately to each region.

Region I. The radial depth is less than the corner radius of the tool (i.e., $0 < y < r$). The chip thickness is constant and equal to the feed rate (i.e., $h = c$). The cutting forces in x, y, z coordinates are parallel to the oblique cutting forces F_t, F_r, F_f respectively, and they are given as follows:

$$\left.\begin{aligned}
F_{x\mathrm{I}} &= F_{t\mathrm{I}} = K_{tc}c(a-r) + K_{te}(a-r), \\
F_{y\mathrm{I}} &= F_{r\mathrm{I}} = K_{rc}c(a-r) + K_{re}(a-r), \\
F_{z\mathrm{I}} &= F_{f\mathrm{I}} = K_{fc}c(a-r) + K_{fe}(a-r).
\end{aligned}\right\} \tag{2.64}$$

The cutting constants (K_{tc}, K_{rc}, K_{fc}) are evaluated from Eqs. (2.62) by using the orthogonal cutting parameters (ϕ_n, τ_s, β_a) obtained from orthogonal cutting tests. However, because the tool has both side and back rake angles, the equivalent oblique angle (i) and normal rake angle (α_n) must be evaluated from Eq. (2.63). The normal friction angle is evaluated from

$$\beta_n = \tan^{-1}(\tan \beta_a \cos i). \tag{2.65}$$

Note that the side cutting angle (or approach angle) is zero for this particular tool in region I. The edge force K_{re} can be taken equal to zero since it does not theoretically exist in the orthogonal cutting test measurements.

Region II. In this region the chip thickness reduces continuously and the oblique cutting forces change their directions around the curved chip segment. The most accurate evaluation can be handled by dividing the chip into small differential elements with an angular increment $d\theta$. The center of the chip's outer surface curvature is O, while the center of its inner curvature is O''. The tool nose radius is r, and the total angular contact is $\angle COB = \theta_0$. The area of differential (dA) chip can be approximated by

$$dA \approx \overline{AT} dS, \tag{2.66}$$

where $dS = rd\theta$, $\overline{AT} = \overline{AO} - \overline{TO}$, $\overline{AO} = r$, $\overline{OO''} = c$, $\overline{TO''} = r$, and

$$\overline{TO} = \sqrt{c^2 + r^2 - 2cr \cos \gamma}. \tag{2.67}$$

Using the law of sines,

$$\frac{\overline{OO''}}{\sin[\pi - (\pi - \theta + \gamma)]} = \frac{\overline{TO''}}{\sin(\pi - \theta)},$$

the following relationship is obtained:

$$\gamma = \theta - \sin^{-1}\left[\frac{c}{r} \sin(\pi - \theta)\right]. \tag{2.68}$$

The instantaneous chip thickness at the position defined by angle θ is given by

$$\overline{AT} = h(\theta) = r - \sqrt{c^2 + r^2 - 2cr \cos \gamma}, \tag{2.69}$$

and the corresponding differential chip area is

$$dA_i = h(\theta)rd\theta. \tag{2.70}$$

The tangential $(dF_{t,II})$, radial $(dF_{r,II})$, and feed $(F_{f,II})$ forces acting on the differential chip element are given by

$$\left.\begin{array}{l} dF_{t,II} = K_{tc}(\theta)dA + K_{te}dS = [K_{tc}(\theta)h(\theta) + K_{te}]rd\theta, \\ dF_{r,II} = K_{rc}(\theta)dA + K_{re}dS = [K_{rc}(\theta)h(\theta) + K_{re}]rd\theta, \\ dF_{f,II} = K_{fc}(\theta)dA + K_{fe}dS = [K_{fc}(\theta)h(\theta) + K_{fe}]rd\theta. \end{array}\right\} \tag{2.71}$$

Because the oblique geometry varies as a function of approach angle θ, the cutting constants at each differential element must be evaluated for an accurate prediction of the cutting forces. Substituting the back rake, side rake, and instantaneous approach angle $(\psi_r = \theta)$ in Eq. (2.63), we can identify the equivalent oblique angle (i) and normal rake angle α_n, which are then used in evaluating the oblique cutting constants given in Eq. (2.62). The edge forces are considered to be the same as those found from orthogonal cutting tests, and the radial component of the edge force is assumed to be zero (i.e., $K_{re} = 0$). The differential oblique cutting forces can be resolved in the x, y, z directions of the lathe, in which the forces can be measured experimentally using a dynamometer:

$$dF_{x,\text{II}} = dF_{\text{t,II}},$$
$$\left.\begin{array}{l} dF_{y,\text{II}} = dF_{\text{f,II}} \cos\theta - dF_{\text{r,II}} \cos\theta, \\ dF_{z,\text{II}} = dF_{\text{f,II}} \cos\theta + dF_{\text{r,II}} \sin\theta. \end{array}\right\} \qquad (2.72)$$

Integrating the differential cutting forces along the curved chip segment gives the total cutting forces contributed by the chip in region II:

$$F_{q,II} = \int_0^{\theta_0} dF_{q,\text{II}}, \quad q = x, y, z, \qquad (2.73)$$

where the approach angle limit is $\theta_0 = \pi - \cos^{-1}(c/2r)$. Because the cutting constants and chip thickness are functions of the instantaneous approach angle θ, the continuous integration in Eq. (2.73) is not trivial. It is much more practical to use constant average cutting coefficients by assuming a mean approach angle of $\pi/2$ in Eqs. (2.62) and (2.63) or by digitally integrating the forces along the curved chip segment. The curved chip is divided into $K = \theta_0/\Delta\theta$ small chip segments, and the cutting forces contributed by each small segment is evaluated and summed numerically:

$$F_{q,\text{II}} = \sum_{k=0}^{K} dF_{q,\text{II}}, \quad q = x, y, z, \qquad (2.74)$$

where the instantaneous approach angle for each segment is $\theta_k = k\Delta\theta$.

The total cutting forces acting on the tool are evaluated by summing the forces generated in regions I and II. Thus

$$F_q = F_{q,\text{I}} + F_{q,\text{II}}, \quad q = x, y, z. \qquad (2.75)$$

The torque (T) and power (P) required from the spindle drive can be predicted by

$$T = F_x\left(\frac{d-a}{2}\right), \quad P = F_x V, \qquad (2.76)$$

where d is the diameter of the shaft in [m] and V is the cutting speed in [m/s]. The units of torque and power are [Nm] and [W], respectively. It is also important to select a suitable feed rate, which does not leave larger feed marks on the surface than specified by the tolerance of the part. The amplitude of the feed mark (R_s) can be evaluated from Figure 2.13 as

$$R_\text{s} = r\left(1 - \cos\left(\sin^{-1}\left(\frac{c}{2r}\right)\right)\right). \qquad (2.77)$$

Similar analysis can be applied to other turning tools by dividing their cutting edges into small oblique elements as evaluated here. The prediction of cutting forces, torque, power, and cutting constants are important in sizing machine tools for a particular operation or for selecting cutting speeds, feeds, and depth of cuts to avoid tool breakage and chatter vibrations. It is equally important to predict the direction of the chip flow and the type of chip produced

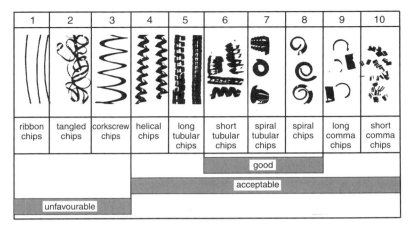

1	2	3	4	5	6	7	8	9	10
ribbon chips	tangled chips	corkscrew chips	helical chips	long tubular chips	short tubular chips	spiral tubular chips	spiral chips	long comma chips	short comma chips

Figure 2.14: Classification of chip forms. Source: Kluft et al. CIRP [25].

by the particular tool geometry used at certain cutting conditions [24]. The chips produced in continuous turning processes may be of a ribbon, tangled, or corkscrew type, which are unfavorable (see Fig. 2.14 [25]). They can rub the finished workpiece or entangle around the tool, and it is hard to remove them from the machine tool with mechanized chip conveyors. Such unfavorable chips also cause danger for the operator's safety; productivity of the machining operation suffers from the scratched finished workpiece surfaces; and the chips can cause tool breakage as they can become entangled around the cutting edge. The basic factors that influence the chip form are the workpiece material, the tool geometry, the cutting fluid, the dynamic properties of the machine tool, and the cutting conditions. Chip breakers, which are clamped on the rake face of the tool or formed on inserts, are used to break long chips. Chip breakers disturb the free flow direction of the chip; they force the chip to curl toward the workpiece or tool, thus creating tensile stresses in the chip that lead to its breakage.

2.7 MECHANICS OF MILLING PROCESSES

A sample vertical CNC machining center, which is capable of milling, drilling, and tapping, is shown in Figure 2.15. The milling operation is an intermittent cutting process using a cutter with one or more teeth. A milling cutter is held in a rotating spindle, while the workpiece clamped on the table is linearly moved toward the cutter. A sample face milling operation and various inserted milling cutters are shown in Figure 2.16. Each milling tooth therefore traces a trochoidal path [26, 27], producing varying but periodic chip thickness at each tooth passing interval. Various milling operations are shown in Figure 2.17. Depending on the workpiece geometry, different milling cutters and machines are used. In this section, the mechanics of the milling process are presented for simple face milling operations. Mechanics of other milling operations are modeled by geometrically extending the mechanics of face milling.

Figure 2.15: A vertical CNC milling machine. Source: Hitachi Seiki Co., Ltd.

The standard tool geometry in milling is shown in Figure 2.18. Double negative tools are shock resistant in heavy-duty face milling operations. Rigid and high power milling machines are suitable for heavy machining with negative cutters. For accurate and light milling operations, double positive milling cutters are ideal. Negative–positive tools produce a good surface finish and are efficient in removing the chips from the insert pockets. There are three types of milling operations used in practice:

• *face milling operations*, in which entry and exit angles of the milling cutter relative to the workpiece are nonzero;

Figure 2.16: A face milling operation and various inserted milling cutters. Source: Mitsubishi Materials.

Figure 2.17: Various milling operations. Source: Mitsubishi Materials.

- ***up-milling operations***, in which the entry angle is zero and the exit angle is nonzero; and

- ***down-milling operations***, in which the entry angle is not zero and the exit angle is zero.

Both up- and down-milling operations are called peripheral or end milling operations. The geometry of chip formation in milling is shown in Figure 2.19. Unlike in turning processes, in milling the instantaneous chip thickness (h) varies periodically as a function of time-varying immersion. The chip thickness variation can be approximated as

$$h(\phi) = c \sin \phi, \tag{2.78}$$

where c is the feed rate (mm/rev-tooth) and ϕ is the instantaneous angle of immersion. First, the helix angle is considered to be zero, which is the case in face milling operations with inserted cutters. Tangential ($F_t(\phi)$), radial ($F_r(\phi)$), and axial ($F_a(\phi)$) cutting forces are expressed as a function of varying uncut chip area ($ah(\phi)$) and edge contact length (a):

$$\left. \begin{aligned} F_t(\phi) &= K_{tc}ah(\phi) + K_{te}a, \\ F_r(\phi) &= K_{rc}ah(\phi) + K_{re}a, \\ F_a(\phi) &= K_{ac}ah(\phi) + K_{ae}a, \end{aligned} \right\} \tag{2.79}$$

Figure 2.18: Standard face milling cutter geometry.

where K_{tc}, K_{rc}, and K_{ac} are the cutting force coefficients contributed by the shearing action in tangential, radial, and axial directions, respectively, and K_{te}, K_{re}, and K_{ae} are the edge constants. If we assume zero nose radius and zero approach angle on the inserts, the axial components of the cutting forces become zero ($F_a = 0$). Otherwise their influence must be modeled as presented in the section on turning (Section 2.6). The cutting coefficients are assumed to be constant for a tool–work material pair, and they can be evaluated either mechanistically from milling tests or using the classical oblique cutting transformations given by Eqs. (2.62) and (2.63). They are sometimes expressed as a nonlinear function (Eq. 2.44) of the instantaneous or average chip thickness h_a [28]. The average chip thickness per revolution is calculated from the swept

Figure 2.19: Geometry of milling process.

zone as

$$h_{\mathrm{a}} = \frac{\int_{\phi_{\mathrm{st}}}^{\phi_{\mathrm{ex}}} c \sin \phi \, d\phi}{\phi_{\mathrm{ex}} - \phi_{\mathrm{st}}} = -c \frac{\cos \phi_{\mathrm{ex}} - \cos \phi_{\mathrm{st}}}{\phi_{\mathrm{ex}} - \phi_{\mathrm{st}}}. \tag{2.80}$$

The instantaneous cutting torque (T_{c}) on the spindle is

$$T_{\mathrm{c}} = F_{\mathrm{t}} \cdot \frac{D}{2}, \tag{2.81}$$

where D is the diameter of the milling cutter. Horizontal (i.e., feed), normal, and axial components of the cutting forces acting on the cutter are derived from the equilibrium diagram shown in Figure 2.19:

$$\left.\begin{aligned} F_x(\phi) &= -F_{\mathrm{t}} \cos \phi - F_{\mathrm{r}} \sin \phi, \\ F_y(\phi) &= +F_{\mathrm{t}} \sin \phi - F_{\mathrm{r}} \cos \phi, \\ F_z(\phi) &= +F_{\mathrm{a}}. \end{aligned}\right\} \tag{2.82}$$

It must be noted that the cutting forces are produced only when the cutting tool is in the cutting zone, that is,

$$F_x(\phi),\ F_y(\phi),\ F_z(\phi)\ > 0 \ \text{ when } \ \phi_{\mathrm{st}} \le \phi \le \phi_{\mathrm{ex}},$$

where ϕ_{st} and ϕ_{ex} are the cutter entry and exit angles, respectively. Another important point is that there may be more than one tooth cutting simultaneously depending on the number of teeth on the cutter and the radial width of cut. The tooth spacing angle ϕ_p (or cutter pitch angle) is given as

$$\phi_p = \frac{2\pi}{N},$$

where N is the number of teeth on the cutter. There will be more than one tooth cutting simultaneously when the swept angle ($\phi_s = \phi_{ex} - \phi_{st}$) is larger than the cutter pitch angle (i.e., $\phi_s > \phi_{ex} - \phi_{st}$). When more than one tooth cuts simultaneously, the contribution of each tooth to total feed and normal forces must be considered. It must also be noted that since each tooth will be away from its neighboring tooth by the amount of pitch angle, the uncut chip thickness removed by each cutting edge will be different at an instantaneous position of the cutter. We can formulate the total feed, normal, and axial forces as

$$F_x = \sum_{j=1}^{N} F_{xj}(\phi_j), \quad F_y = \sum_{j=1}^{N} F_{yj}(\phi_j), \quad F_z = \sum_{j=1}^{N} F_{zj}(\phi_j), \tag{2.83}$$

whenever $\phi_{st} \leq \phi_j \leq \phi_{ex}$. Each term in the summation block represents the contribution of each tooth to the cutting forces. If the tooth j is out of the immersion zone, it contributes zero to total forces. The instantaneous resultant cutting force on the cutter (or workpiece) is given as

$$F = \sqrt{F_x^2 + F_y^2 + F_z^2}. \tag{2.84}$$

Instantaneous cutting torque on the spindle is

$$T_c = \frac{D}{2} \cdot \sum_{j=1}^{N} F_{tj}(\phi_j) \longrightarrow \phi_{st} \leq \phi_j \leq \phi_{ex}, \tag{2.85}$$

where D is the diameter of the cutter. The cutting power (P_t) drawn from the spindle motor is

$$P_t = V \cdot \sum_{j=1}^{N} F_{tj}(\phi_j) \longrightarrow \phi_{st} \leq \phi_j \leq \phi_{ex}, \tag{2.86}$$

where $V = \pi D n$ is the cutting speed and n is the spindle speed. For a given set of cutting conditions, the engineer may be required to predict the maximum cutting power, torque, and cutting forces required from the machine tool spindle and feed drives. The cutting forces, torque, and power are uniformly periodic at tooth passing frequency. Periodic cutting forces dynamically load and unload the machine tool structure, workpiece, and the cutter at each tooth period. Typical resultant cutting forces for three types of milling operations are shown in Figure 2.20. Half-immersion (i.e., $b = D/2$) up- and down-milling forces have opposite trends. The chip load starts with zero and gradually increases to maximum at the exit in up-milling; hence forces have the same trend. However, the tooth experiences maximum chip load during entry followed by a gradual decrease of the chip load and hence the cutting forces. Manufacturing

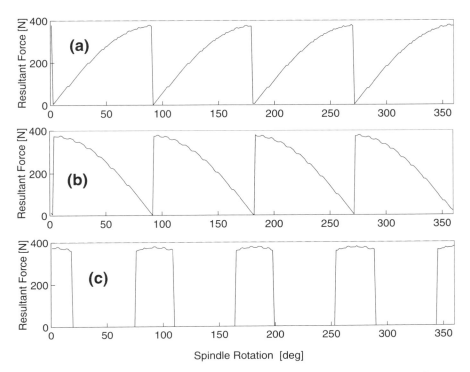

Figure 2.20: Simulated resultant milling forces. Face milling cutter: $N = 4$ teeth, $a = 2$ mm, $c = 0.1$ mm/tooth, $K_t = 1,800$ MPa, $K_r = 0.3$. a) Half immersion up milling, b) half immersion down milling, c) center face milling with $\phi_{st} = 75°$, $\phi_{ext} = 105°$.

engineers are advised to use up-milling operations for heavy metal removal rates where the shock loading is reduced. For light finish cuts, down-milling is preferred in order to obtain a smooth surface finish. Center face milling has severe interrupted, pulse loading of the machine tool, which is not advised for light machines and positive tools. Pulse loading of the machine may resonate various structural vibration modes as well as cause transient vibrations during each entry and exit.

2.7.1 Mechanics of Helical End Mills

Periodic loading causes cyclic mechanical and thermal stresses on the tool, which leads to a shorter tool life. Helical end mills are used to dampen the sharp variations in the oscillatory components of the milling forces, and they are used when the depth of cut is large but the width of cut is small. Their primary function is peripheral milling, where the walls of parts are the target finish surface. A typical end milling cutter with helical flutes is shown in Figure 2.21. The helix on the cutter provides a gradually increasing chip load along the helical flutes of the end mill [29]. If the helix angle on the cutter is β, a point on the axis of the cutting edge will be lagging behind the end point of the tool. The lag angle (ψ) at the axial depth of cut (z) is found as

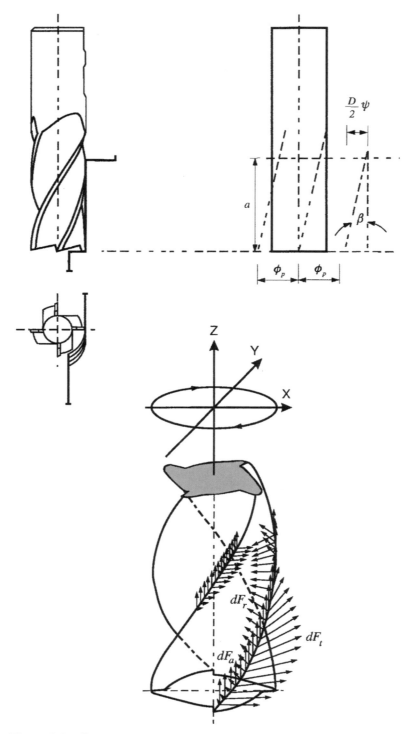

Figure 2.21: Geometry of helical end milling.

(Fig. 2.21)

$$\tan \beta = \frac{D\psi}{2z} \quad \text{and}$$

$$\psi = \frac{2z \tan \beta}{D}. \quad (2.87)$$

a) Half immersion up milling with a cylindrical end mill

When the bottom point of a reference flute of the end mill is at immersion angle ϕ, a cutting edge point that is axially z [mm] above will have an immersion angle of $(\phi - \psi)$. Obviously, the chip thickness removed along the flute's axis will also be different at each point. A general pseudocode of a milling force simulation program is given in Table 2.2. The input variables set by the user are helix, entry, and exit angles, axial depth of cut, the number of teeth, feed rate, spindle speed, cutter diameter, and cutting constants. The cutter is rotated with small incremental angles. At each incremental rotation,

b) Slot ball end milling

Figure 2.22: Simulated and measured cutting forces in end milling of Ti$_6$Al$_4$V alloy. Feed rate = 0.05 mm/tooth, $\alpha_n = 0$. a) $R_0 = 9.05$ mm, $V = 30$ m/min, $N_f = 4$ flutes, $\alpha_n = 12°$, axial depth of cut = 5.08 mm, $i_0 = 30°$. b) Single flute ball end mill with radius $R_0 = 9.525$ mm, nominal helix angle $i_0 = 30°$, axial depth of cut = 1.27 mm, spindle speed $N = 269$ rev/min.

the cutting forces are integrated axially along the sliced differential elements from the bottom of the flute toward the final axial depth of cut.

Sample experimental and simulation results for helical cylindrical and helical ball end milling of a titanium alloy are given in Figure 2.22 [18, 30]. The cutting coefficients are obtained by transforming orthogonal cutting constants of titanium given in Table 2.1 to oblique end mill geometry (see Eq. 2.62). The agreement between the experimental and simulation results are very satisfactory owing to careful evaluation of shear stress, shear angle, and friction coefficient in a series of orthogonal tests, as well as accurate modeling of geometry and oblique transformation of mechanics. More complex milling cutters and operations can also be modeled by employing the same mechanics approach, by designing a generalized, parametric milling cutter geometry as presented by Altintas and Lee [30].

2.8 ANALYTICAL MODELING OF END MILLING FORCES

The discrete simulation of cutting forces in end milling has been explained in the previous section. The accuracy of the cutting force prediction strongly depends on the selected digital integration interval. When the axial depth of cut is large

TABLE 2.2. Pseudocode for Milling Force Simulation Algorithm

; **Inputs:**

Cutting conditions	$:a, c, n, \phi_{st}, \phi_{ex}$
Tool geometry	$:D, N, \beta$
Cutting constants	$:K_{tc}, K_{rc}, K_{te}, K_{re}$
Integration angle	$:\Delta\phi$
Integration height	$:\Delta a$

; **Outputs:**

Cutting force history	$:F_x(\phi), F_y(\phi), F(\phi)$
Cutting torque and power history	$:T_c(\phi), P_c(\phi)$

;

; **Variables**

;

$\phi_p = \frac{2\pi}{N}$;Cutter pitch angle

$K = \frac{2\pi}{\Delta\phi}$;Number of angular integration steps

$L = \frac{a}{\Delta a}$;Number of axial integration steps

$i = 1$ to K ;Angular integration loop

   $\phi(i) = \phi_{st} + i\Delta\phi$;Immersion angle of flute's bottom edge

   $F_x(i) = F_y(i) = F_t(i) = 0.0$;Initialize the force integration registers

   $k = 1$ to N ;Calculate the force contributions of all teeth

   $\phi_1 = \phi(i) + (k-1)\phi_p$;Immersion angle for tooth k

   $\phi_2 = \phi_1$;Memorize the present immersion

;

   $j = 1$ to L ;Integrate along the axial depth of cut

   $a(j) = j \cdot \Delta a$;Axial position

   $\phi_2 = \phi_1 - \frac{2\tan\beta}{D}a(j)$;Update the immersion angle due helix

   if $\phi_{st} \le \phi_2 \le \phi_{ex}$;if the edge is cutting
                                    ;then

   $h = c\sin\phi_2$;Chip thickness at this point

   $\Delta F_t = \Delta a(K_{tc}h + K_{te})$;Differential tangential force

   $\Delta F_r = \Delta a(K_{rc}h + K_{re})$;Differential radial force

   $\Delta F_x = -\Delta F_t\cos\phi_2 - \Delta F_r\sin\phi_2$;Differential feed force

   $\Delta F_y = \Delta F_t\sin\phi_2 - \Delta F_r\cos\phi_2$;Differential normal force

;

   $F_x(i) = F_x(i) + \Delta F_x$;Sum the cutting forces

   $F_y(i) = F_y(i) + \Delta F_y$;contributed by the all

   $F_t(i) = F_t(i) + \Delta F_t$ 'active edges

   else

   next j

   next k

;

; Resulting cutting force values at immersion angle $\phi(i)$

;

   $F(i) = \sqrt{F_x^2(i) + F_y^2(i)}$;Resultant cutting force

   $T_c(i) = \frac{D}{2}F_t(i)$;Cutting torque

next i

;

; Plot $F_x(i), F_y(i), F_i, T_c(i)$ with varying immersion $\phi(i)$

;

stop

end

in helical end milling operations, the differential element height in the axial direction must be very small to avoid numerical oscillations on the cutting force wave forms. When the cutting forces are used to predict the vibrations of the end mill or workpiece, the numerical oscillations lead to faulty simulation of vibrations. Also, an accurate prediction of force distribution along the end mill and flexible thin webs is necessary to predict the dimensional form errors left on the finish surface. When kinematics and certain properties of the milling process are considered, it is possible to derive semianalytical expressions for end milling forces [31]. An end mill with a helix angle of β, diameter of D, and N number of flutes is assumed (see Fig. 2.21). The axial depth of cut (a) is constant and the immersion is measured clockwise from the normal (y) axis. Assuming that the bottom end of one flute is designated as the reference immersion angle ϕ, the bottom end points of the remaining flutes are at angles $\phi_j(0) = \phi + j\phi_p$; $j = 0, 1, 2 \ldots (N-1)$. At an axial depth of cut z the lag angle is $\psi = k_\beta z$, where $k_\beta = (2 \tan \beta)/D$. The immersion angle for flute j at axial depth of cut z therefore is

$$\phi_j(z) = \phi + j\phi_p - k_\beta z. \tag{2.88}$$

Tangential $(dF_{t,j})$, radial $(dF_{r,j})$, and axial $(dF_{a,j})$ forces acting on a differential flute element with height dz are expressed similar to Eq. (2.79):

$$\left.\begin{aligned}
dF_{t,j}(\phi, z) &= [K_{tc}h_j(\phi_j(z)) + K_{te}]dz, \\
dF_{r,j}(\phi, z) &= [K_{rc}h_j(\phi_j(z)) + K_{re}]dz, \\
dF_{a,j}(\phi, z) &= [K_{ac}h_j(\phi_j(z)) + K_{ae}]dz,
\end{aligned}\right\} \tag{2.89}$$

where the chip thickness is

$$h_j(\phi, z) = c \sin \phi_j(z). \tag{2.90}$$

The directions of the cutting forces are aligned with the cutter axis. The cutting constants can be evaluated from Eq. (2.62) by accepting the helix angle as the oblique angle of the end mill (i.e., $i = \beta$). The elemental forces are resolved into feed (x), normal (y), and axial (z) directions using the transformation

$$\begin{aligned}
dF_{x,j}(\phi_j(z)) &= -dF_{t,j} \cos \phi_j(z) - dF_{r,j} \sin \phi_j(z), \\
dF_{y,j}(\phi_j(z)) &= +dF_{t,j} \sin \phi_j(z) - dF_{r,j} \cos \phi_j(z), \\
dF_{z,j}(\phi_j(z)) &= +dF_{a,j}.
\end{aligned} \tag{2.91}$$

Substituting the differential forces (Eq. 2.89) and the chip thickness (Eq. 2.90) into Eq. (2.91) leads to

$$\left.\begin{aligned}
dF_{x,j}(\phi_j(z)) &= \left\{ \frac{c}{2}[-K_{tc} \sin 2\phi_j(z) - K_{rc}(1 - \cos 2\phi_j(z))] \right. \\
&\quad \left. + [-K_{te} \cos \phi_j(z) - K_{re} \sin \phi_j(z)] \right\}dz, \\
dF_{y,j}(\phi_j(z)) &= \left\{ \frac{c}{2}[K_{tc}(1 - \cos 2\phi_j(z)) - K_{rc} \sin 2\phi_j(z)] \right. \\
&\quad \left. + [K_{te} \sin \phi_j(z) - K_{re} \cos \phi_j(z)] \right\}dz, \\
dF_{z,j}(\phi_j(z)) &= [K_{ac}c \sin \phi_j(z) + K_{ae}]dz.
\end{aligned}\right\} \tag{2.92}$$

The differential cutting forces are integrated analytically along the in-cut portion of the flute j in order to obtain the total cutting force produced by the flute:

$$F_q(\phi_j(z)) = \int_{z_{j,1}}^{z_{j,2}} dF_q(\phi_j(z))dz, \quad q = x, y, z, \tag{2.93}$$

where $z_{j,1}(\phi_j(z))$ and $z_{j,2}(\phi_j(z))$ are the lower and upper axial engagement limits of the in-cut portion of the flute j. The integrations are carried out by noting $\phi_j(z) = \phi + j\phi_p - k_\beta z$, $d\phi_j(z) = -k_\beta dz$. Thus

$$\left.\begin{aligned}
F_{x,j}(\phi_j(z)) &= \left\{ \frac{c}{4k_\beta}\left[-K_{tc}\cos 2\phi_j(z) + K_{rc}[2\phi_j(z) - \sin 2\phi_j(z)]\right] \right. \\
&\left. \quad + \frac{1}{k_\beta}[K_{te}\sin\phi_j(z) - K_{re}\cos\phi_j(z)]\right\}_{z_{j,1}(\phi_j(z))}^{z_{j,2}(\phi_j(z))}, \\
F_{y,j}(\phi_j(z)) &= \left\{ \frac{-c}{4k_\beta}\left[K_{tc}(2\phi_j(z) - \sin 2\phi_j(z)) + K_{rc}\cos 2\phi_j(z)]\right] \right. \\
&\left. \quad + \frac{1}{k_\beta}[K_{te}\cos\phi_j(z) + K_{re}\sin\phi_j(z)]\right\}_{z_{j,1}(\phi_j(z))}^{z_{j,2}(\phi_j(z))}, \\
F_{z,j}(\phi_j(z)) &= \frac{1}{k_\beta}[K_{ac}c\cos\phi_j(z) - K_{ae}\phi_j(z)]_{z_{j,1}(\phi_j(z))}^{z_{j,2}(\phi_j(z))}.
\end{aligned}\right\} \tag{2.94}$$

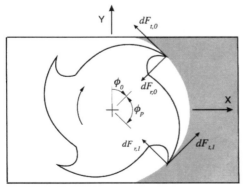

Figure 2.23: Helical flute-part face integration zones, and differential forces on the cutting tool at a particular rotation angle ϕ and depth z.

The axial integration limits $z_{j,1}$ and $z_{j,2}$ are required for each flute to implement the cutting force model. When extruded axially each $[\phi_{st}, \phi_{ex}]$ cutter arc segment defines an immersion section that can interact with a helical tooth j in five distinct ways (Fig. 2.23). The lag angle at full axial depth of cut $z = a$ is $\psi_a = k_\beta a$.

The following computer algorithm is used in determining the axial integration boundaries:

- If $\phi_{st} < \phi_j(z = 0) < \phi_{ex}$ then $z_{j,1} = 0$;

  Case 0: If $\phi_{st} < \phi_j(z = a) < \phi_{ex}$ then $z_{j,2} = a$.

  Case 1: $\phi_j(z = a) < \phi_{st}$ then $z_{j,2} = (1/k_\beta)(\phi + j\phi_p - \phi_{st})$.

- If $\phi_j(z = 0) > \phi_{ex}$ and $\phi_j(z = a) < \phi_{ex}$ then $z_{j,1} = (1/k_\beta)(\phi + j\phi_p - \phi_{ex})$;

  Case 2: If $\phi_j(z = a) > \phi_{st}$ then $z_{j,2} = a$.

  Case 3: If $\phi_j(z = a) < \phi_{st}$ then $z_{j,2} = (1/k_\beta)(\phi + j\phi_p - \phi_{st})$.

  Case 4: If $\phi_j(z = 0) > \phi_{ex}$ and $\phi_j(z = a) > \phi_{ex}$ the flute is out of cut.

The values for the integration limits $z_{j,1}$ and $z_{j,2}$ are taken from the cases listed above substituted into (2.94), and the resulting expressions can be further simplified for an efficient computation in computer programs. Note that to use the expressions, flute $j = 0$ must be aligned at $\phi = 0$ in the beginning of the algorithm. The remaining flutes must be indexed ($j = 1, 2, \ldots, N-1$) from the reference tooth at pitch angle (ϕ_p) intervals. The cutting forces contributed by all flutes are calculated and summed to obtain the total instantaneous forces on the cutter at immersion ϕ:

$$F_x(\phi) = \sum_{j=0}^{N-1} F_{x_j}; \quad F_y(\phi) = \sum_{j=0}^{N-1} F_{y_j}; \quad F_z(\phi) = \sum_{j=0}^{N-1} F_{z_j}. \tag{2.95}$$

The resultant cutting force acting on the milling cutter is

$$F(\phi) = \sqrt{F_x(\phi)^2 + F_y(\phi)^2 + F_z(\phi)^2}. \tag{2.96}$$

Thus computationally inefficient and approximate digital integration methods are eliminated by the closed form expressions of the instantaneous cutting forces. The closed form expressions can be used for process planning, investigating the interaction between tool–workpiece structure and the milling process, and predicting finish surface. The algorithm can be efficiently implemented to the CAD/CAM systems for milling process simulation [32].

2.8.1 Mechanistic Identification of Cutting Constants in Milling

Using orthogonal cutting parameters such as shear angle, shear stress, and friction coefficient to determine oblique milling constants is desired for modeling a variety of milling cutter geometries (see Eqs. 2.62 and 2.62). However, some cutting tools may have complex cutting edges, and the evaluation of cutting constants by creating a time-consuming orthogonal cutting data base may not be possible. In such cases, a quick method of calibrating the milling tools, the mechanistic approach, is used [18]. A set of milling experiments are conducted at different feed rates but constant immersion and axial depth of cut. The average forces per tooth period are measured. To avoid the influence of run out on the measurements, the total force per spindle revolution is collected and divided by the number of teeth on the cutter. The experimentally evaluated average cutting forces are equated to analytically derived average milling force expressions, which leads to the identification of cutting constants. Since the total material removed per tooth period is constant with or without helix angle, the average cutting forces are independent of helix angle. Replacing $dz = a$, $\phi_j(z) = \phi$, and $k_\beta = 0$ in Eqs. (2.92) and integrating them over one revolution and dividing by the pitch angle ($\phi_p = 2\pi/N$) yields to average milling forces per tooth period,

$$\overline{F}_q = \frac{1}{\phi_p} \int_{\phi_{st}}^{\phi_{ex}} F_q(\phi)d\phi, \tag{2.97}$$

since the flute cuts only within the immersion zone (i.e., $\phi_{st} \leq \phi \leq \phi_{ex}$).

Integrating the instantaneous cutting forces leads to

$$
\begin{aligned}
\overline{F}_x &= \left\{ \frac{Nac}{8\pi} \left[K_{\mathrm{tc}} \cos 2\phi - K_{\mathrm{rc}}[2\phi - \sin 2\phi] \right] \right. \\
&\quad \left. + \frac{Na}{2\pi}[-K_{\mathrm{te}} \sin \phi + K_{\mathrm{re}} \cos \phi] \right\}_{\phi_{\mathrm{st}}}^{\phi_{\mathrm{ex}}}, \\
\overline{F}_y &= \left\{ \frac{Nac}{8\pi} \left[K_{\mathrm{tc}}(2\phi - \sin 2\phi) + K_{\mathrm{rc}} \cos 2\phi \right] \right. \\
&\quad \left. - \frac{Na}{2\pi}[K_{\mathrm{te}} \cos \phi + K_{\mathrm{re}} \sin \phi] \right\}_{\phi_{\mathrm{st}}}^{\phi_{\mathrm{ex}}}, \\
\overline{F}_z &= \frac{Na}{2\pi}[-K_{\mathrm{ac}}c \cos \phi + K_{\mathrm{ae}}\phi]_{\phi_{\mathrm{st}}}^{\phi_{\mathrm{ex}}}.
\end{aligned} \tag{2.98}
$$

Full immersion (i.e., slotting) milling experiments are most convenient; here the entry and exit angles are $\phi_{\mathrm{st}} = 0$ and $\phi_{\mathrm{ex}} = \pi$ respectively. When full immersion conditions are applied to Eqs. (2.98), the average forces per tooth period are simplified as

$$
\begin{aligned}
\overline{F}_x &= -\frac{Na}{4} K_{\mathrm{rc}}c - \frac{Na}{\pi} K_{\mathrm{re}}, \\
\overline{F}_y &= +\frac{Na}{4} K_{\mathrm{tc}}c + \frac{Na}{\pi} K_{\mathrm{te}}, \\
\overline{F}_z &= +\frac{Na}{\pi} K_{\mathrm{ac}}c + \frac{Na}{2} K_{\mathrm{ae}}.
\end{aligned} \tag{2.99}
$$

The average cutting forces can be expressed by a linear function of feed rate (c) and an offset contributed by the edge forces:

$$
\overline{F}_q = \overline{F}_{qc}c + \overline{F}_{qe} \quad (q = x, y, z). \tag{2.100}
$$

The average forces at each feed rate are measured, and the cutting-edge components ($\overline{F}_{qc}, \overline{F}_{qe}$) are estimated by a linear regression of the data. Finally, the cutting force coefficients are evaluated from Eqs. (2.99) and (2.100) as follows:

$$
\begin{aligned}
K_{\mathrm{tc}} &= \frac{4\overline{F}_{yc}}{Na}, \quad K_{\mathrm{te}} = \frac{\pi \overline{F}_{ye}}{Na}, \\
K_{\mathrm{rc}} &= \frac{-4\overline{F}_{xc}}{Na}, \quad K_{\mathrm{re}} = \frac{-\pi \overline{F}_{xe}}{Na}, \\
K_{\mathrm{ac}} &= \frac{\pi \overline{F}_{zc}}{Na}, \quad K_{\mathrm{ae}} = \frac{2\overline{F}_{ze}}{Na}.
\end{aligned} \tag{2.101}
$$

The procedure is repeated for each cutter geometry; hence the milling force coefficients can not be predicted prior to testing of newly designed cutters using mechanistic models. However, oblique cutting transformation using basic orthogonal cutting parameters can predict the cutting constants before the cutter is manufactured.

2.9 MECHANICS OF DRILLING

A sample twist drill, which is used in drilling holes, is shown in Figure 2.24. A twist drill has a chisel edge at the bottom and two helical cutting lips with a taper angle (κ_t) that meet with the flutes with a helix angle of β_0. The helical flutes do not cut; they are used to evacuate the chips from the drilled hole. The chisel has a width of $2w$ and an edge angle of ψ_c. The cutting lips have an offset from the drill center due to chisel. The lips expand the hole by removing the material with a constant chip thickness (h) as the drill is fed into the material at a feed rate of c [mm/rev]. The thrust force that is used to push the drill into the work material, and the torque applied to the drill and spindle drive, are required to evaluate the mechanics of the drilling operations. The mechanics of drilling must be analyzed separately for the chisel and cutting lip regions.

Chisel Edge. The chisel edge does not cut but only spreads the material sideways by an indentation mechanism. Instead of using the laws of cutting, the mechanics of indentation must be used. If the process is simplified as in a hardness test, then the thrust force acting on the chisel edge can be simplified as

$$THRUST_i = F_{z,i} = A_{ch}H_B,\qquad(2.102)$$

where H_B is the Brinell hardness of the work material and A_{ch} is the instantaneous indentation area of the chisel edge. A_{ch} is evaluated as the product of chisel length ($2w/\sin(\pi - \psi_c)$) and the contact length of spread material with the lip ($c/(2\cos\gamma_t)$). Considering both sides of the chisel contact, we have an

Figure 2.24: Twist drill geometry

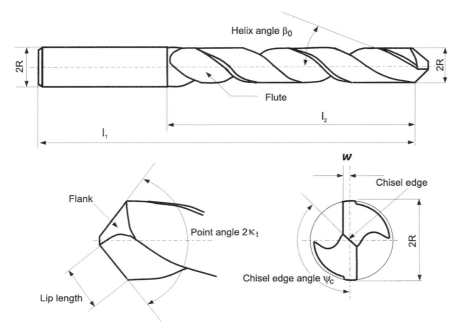

area of contact for indentation of

$$A_{ch} = \frac{2wc}{\sin(\pi - \psi_c)\cos\kappa_t}. \tag{2.103}$$

Approximating the chisel force by the simple indentation method is not reliable for an accurate analysis. The chisel edge geometry and indentation mechanism is rather complex and requires detailed geometric modeling as well as experimental calibration of various empirical factors [33]. Recently, the chisel geometry has been improved significantly by tool manufacturers to minimize skidding of the drill on the work surface during penetration. For practical considerations, the chisel forces can be assumed to be about 10–15% of the lip cutting forces, and the torque can be neglected since the chisel width ($2w$) is rather small.

Cutting Lip. The geometry of the cutting lip of a drill is quite complex. To use the orthogonal to oblique cutting transformation, it is necessary to identify the helix, normal rake, and oblique angles at cutting points along the lip. Because of the offset caused by the chisel and the varying diameter, the helix, normal rake, and oblique angles vary from the chisel–lip to lip–flute intersection. The following treatment of the drill geometry is based on the work presented by Galloway [34] and Armarego [2].

The geometric model of the drill shown in Figure 2.25 is used in explaining the oblique cutting model. The drill axis is aligned with z, the lip that has an offset from the drill axis due to chisel edges is aligned parallel to x, and the axis normal to the lip is parallel to y in a Cartesian coordinate system with a center at the drill tip O.

First consider the bottom of the flute where the lips and chisel edge meet (i.e., the plane of $z = 0$). The chisel web offset from the drill axis is w, and the chisel edge angle is ψ_c (see the detailed view in Fig. 2.26). The radial distance between the drill center and the point where the lip intersects with the chisel edge is

$$r(0) = \frac{w}{\sin(\pi - \psi_c)}, \tag{2.104}$$

with coordinates $x(0) = r(0)\cos(\pi - \psi_c)$, $y(0) = w$, $z = 0$.

The lip and helical flute intersect at elevation $z = a$ and the drill diameter is R (see detail (b) in Fig. 2.26). The radial distance between the lip's outermost point and drill center is $r(a) = R$, with coordinates

$$\left.\begin{array}{l} x(a) = R\cos\theta(a), \quad \theta(a) = \sin^{-1}(w/R), \\ y(a) = w, \\ z = a, \end{array}\right\} \tag{2.105}$$

where R is the drill radius. The projected length of the whole cutting lip on the xy plane at elevation $z = a$ becomes

$$\overline{D'E} = \overline{DE}\sin\kappa_t = b\sin\kappa_t, \tag{2.106}$$

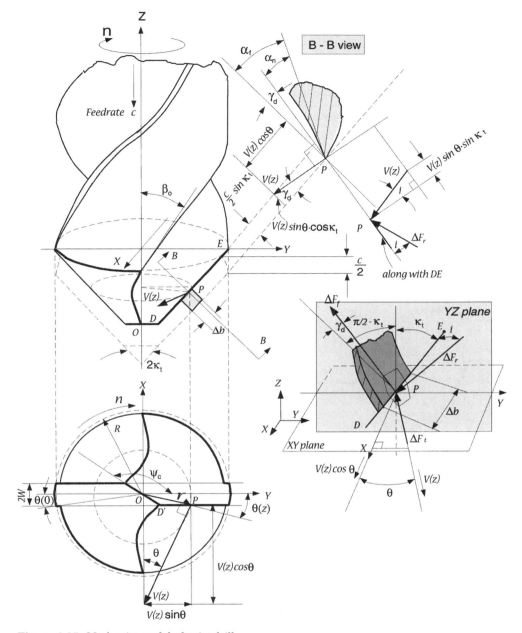

Figure 2.25: Mechanics model of twist drill.

where $b = \overline{DE}$ is the length of lip that cuts the work material. Although the helix angle changes along the lip, a nominal helix angle (β_0) of the drill is defined for the flutes at the cylindrical region ($r(a) = R$):

$$\tan \beta_0 = \frac{2\pi R}{L_\mathrm{p}}, \tag{2.107}$$

where L_p is the constant pitch length of the helix.

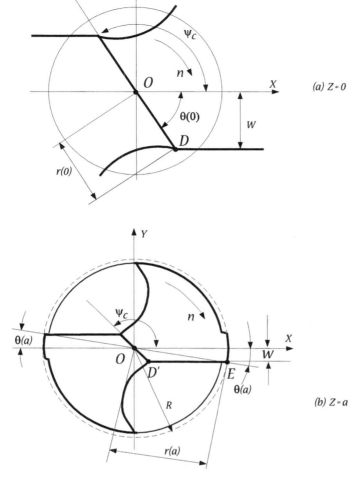

Figure 2.26: Top view of twist drill at the bottom and at the flute–lip intersection planes.

A point $P(x, y, z)$ on the cutting lip is considered in Figure 2.25. The projection of the cutting lip on the xy plane at elevation z is $\overline{D'P} = z \tan \kappa_t$. The radial distance between the axis of the drill and point P is

$$r(z) = \sqrt{y(0)^2 + [x(0) + \overline{D'P}]^2} = \sqrt{w^2 + [w \cot(\pi - \psi_c) + z \tan \kappa_t]^2} \qquad (2.108)$$

with coordinates

$$x(z) = r(z) \cos(\theta(z)), \quad y(z) = r(z) \sin(\theta(z)). \qquad (2.109)$$

The local helix angle at point P can be expressed as

$$\beta(z) = \frac{2\pi r(z)}{L_p}, \qquad (2.110)$$

which has a different value at each elevation z. The cutting velocity (V) is perpendicular to the radius $r(z)$ and has the following components in the xy plane:

$$V_x(z) = V \cdot \cos \theta(z), \quad V_y(z) = V \cdot \sin \theta(z). \tag{2.111}$$

The projection of the cutting velocity on the cutting lip, which has a taper angle of κ_t, is

$$V_t = V_y \sin \kappa_t = V \sin \theta(z) \sin \kappa_t. \tag{2.112}$$

The oblique angle (i) is defined between the cutting velocity and the normal to the cutting edge:

$$\sin i = \frac{V_t}{V} = \sin \theta(z) \sin \kappa_t. \tag{2.113}$$

The cutting mechanics are defined on the normal plane, which is perpendicular to the cutting edge. If a normal plane to cutting edge is considered at point P, the velocity (V) has a component perpendicular to the tapered cutting edge $(V_y(z) \cos \kappa_t)$ and a component parallel to V_y. The angle between the two velocity components is

$$\tan \gamma_d = \frac{V_y \cos \kappa_t}{V_x} = \tan \theta(z) \cos \kappa_t. \tag{2.114}$$

The effective rake angle (α_f) is evaluated by Armarego and Brown by considering a point inside the lip and along the cutting velocity and projecting it on the differential. This yields [2]

$$\tan \alpha_f = \frac{\tan \beta(z) \cos \theta(z)}{\sin \kappa_t - \tan \beta(z) \sin \theta(z) \cos \kappa_t}. \tag{2.115}$$

The normal rake angle is found from the geometry via

$$\alpha_n = \alpha_f - \gamma_d. \tag{2.116}$$

Although the drill geometry is complex, one can take the geometric relationships (2.114–2.116) for granted and predict the cutting forces along the lip. If the drill lip is broken into small differential elements with height dz and width Δb, each differential lip removes a chip with an area of

$$dA(z) = \Delta b \cdot h, \tag{2.117}$$

where the chip thickness (h) removed by one of the two flutes and the chip width (Δb) are

$$h = \frac{c}{2} \sin \kappa_t, \quad \Delta b = \frac{dz}{\cos \kappa_t}. \tag{2.118}$$

The cutting forces in tangential (parallel to the cutting velocity) chip flow and radial directions are expressed as

$$
\left.
\begin{aligned}
dF_t(z) &= K_{tc}(z)dA + K_{te}\Delta b, \\
dF_f(z) &= K_{fc}(z)dA + K_{fe}\Delta b, \\
dF_r(z) &= K_{rc}(z)dA + K_{re}\Delta b,
\end{aligned}
\right\}
\tag{2.119}
$$

where the cutting coefficients for each element at elevation z are different due to varying helix, normal rake, and oblique angles. Edge constants K_{te}, K_{fe}, and K_{re} are evaluated experimentally, and cutting constants K_{tc}, K_{fc}, and K_{re} are evaluated from an orthogonal cutting database using the oblique transformations given by Eqs. (2.62).

The components of the elemental cutting force components (dF_t, dF_f, dF_r) can be evaluated in the x, y, z directions as shown in Figure 2.27:

$$
\left.
\begin{aligned}
dF_x(z) &= [dF_f \sin \gamma_d - dF_t \cos \theta - dF_r \sin i], \\
dF_y(z) &= [dF_r(\sin i \cdot \sin \gamma_d \cdot \cos \kappa_t) - dF_f \cos \gamma_d \cdot \cos \kappa_t - dF_t \sin \theta], \\
dF_z(z) &= [dF_f \cos \gamma_d \cdot \sin \kappa_t - dF_r(\cos i \cdot \cos \kappa_t + \sin i \cdot \sin \gamma_d \cdot \sin \kappa_t)].
\end{aligned}
\right\}
$$

$$\tag{2.120}$$

Figure 2.27: Direction of drilling forces.

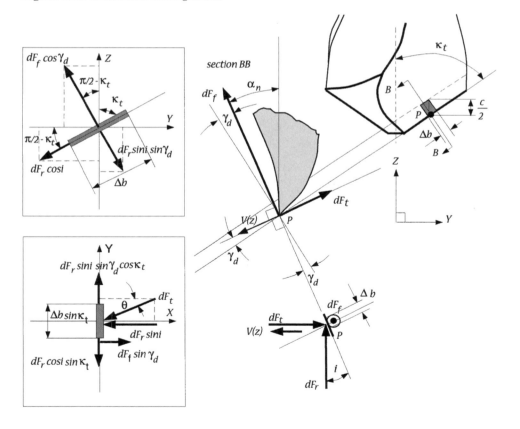

The total thrust and torque exerted on the drill can be evaluated by summing the contribution of all lip elements that number $M = b/\Delta b = b\cos\kappa_t/dz)$. The total thrust force and torque produced by the two lips (i.e., region II) can be expressed as

$$
\left.
\begin{aligned}
THRUST_{ii} &= 2\sum_{m=1}^{M} dF_z(z), \\[2mm]
TORQUE_{ii} &= 2\sum_{m=1}^{M} dF_t(z)\cdot r(z).
\end{aligned}
\right\}
\tag{2.121}
$$

The total thrust force exerted on the drill is found by summing the contributions of the chisel and lip forces:

$$
\text{THRUST} = THRUST_i + THRUST_{ii}.
\tag{2.122}
$$

The torque produced by the chisel edge can be neglected.

The mechanics of chisel and difficulty in modeling the oblique geometry complicate the mechanics of drilling. Readers are referred to a detailed study presented by Armarego et al. [33]. The grinding of twist drills requires accurate modeling of chisel and lip geometry, as well as rake, helix, and clearance angles along the lip and chisel because strongly influence the cutting mechanics [35], vibrations [36], and tool wear [37].

2.10   TOOL WEAR AND TOOL BREAKAGE

Cutting tools can be used only when their edges produce parts with specified surface finish and dimensional tolerance. When the quality of the cutting edge is lost due to wear or breakage (Fig. 2.28), the tool reaches its life limit and must be replaced by a new one. Cutting tools experience various forms of wear and breakage during machining. Tool wear is defined as a gradual loss of tool material at workpiece material and tool contact zones [38]. The moving chip is in contact with the part of the rake face where crater wear occurs (see Fig. 2.29). The cutting edge generates the finish surface, and as it wears, the flank face of the tool starts rubbing against the finish surface leading to flank wear (see Fig. 2.30). A piece of the cutting edge may suddenly fracture during machining due to thermal or mechanical overloading of the cutting edge, increased cutting forces caused by significant tool wear, or weakening of the edge due to thermally or mechanically induced microcracks. The objective of the machining engineer is to select a cutting tool material and geometry, as well as optimal cutting conditions (i.e., feed, speed, depth of cut, and lubricant) that lead to the most economical machining time without violating the surface finish and tolerance of the workpiece and wear and breakage limits of the cutting tool simultaneously. The identification of conditions leading to optimal machining is called a *machinability study*, and its simple fundamentals are briefly introduced in the following.

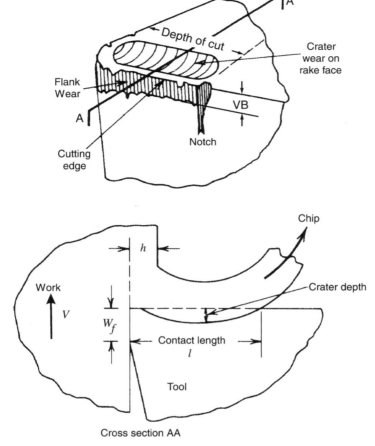

Figure 2.28: The types of tool wear and breakage.

2.10.1 Tool Wear

There are several wear mechanism that may occur simultaneously, or one of them may dominate the process. They can be listed as *abrasion, adhesion, diffusion, fatigue*, and *chemical* wear.

Abrasion Wear

Abrasion occurs when a harder material (i.e., the tool) shears away small particles from the softer work material. However, softer work material also removes small particles from the tool material although at a smaller rate. The hard tool particles are caught between the hard tool and soft work material, and this causes additional abrasion wear. Tool and work materials contain carbides, oxides, and nitrides with hard microstructures; these cause abrasion wear during machining.

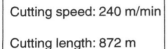

Cutting speed: 240 m/min

Cutting length: 872 m

Cutting speed: 400 m/min

Cutting length: 907 m

Cutting speed: 800 m/min

Cutting length: 780 m

Figure 2.29: Wear of CBN tools in turning P20 mold steel at cutting speeds 240 m/min, 400 m/min, 800 m/min.

Adhesion Wear

When there is a relative motion between the two bodies that are under the normal load, fragments of softer work material adhere to the harder tool. The adhered material is unstable, and it separates from the cutting tool and tears small fragments of the tool material. The typical example in metal cutting is a built-up edge, which usually occurs at low cutting speeds when part of the chip material welds to the cutting edge (see Fig. 2.30). Depending on the size and stability of the built-up edge, either the forces decrease because the effective rake angle becomes positive or the lumped built-up edge dulls the tool and increases the forces. An unstable, large built-up edge occurs close to the cutting edge at low speeds where the tool–chip interface temperature is low (i.e., less than austenitic temperature). The material is still strong at this point and difficult to move over the rake face. As the chip moves over the rake

Figure 2.30: Flank wear and build-up edge.

Flank wear

Build-up edge

face, the chip–tool interface temperature increases, leading to a softer chip, which is easier to move. As the cutting speed is increased, the magnitude and length of the built-up edge becomes smaller and localizes close to the cutting edge. Predicting the tool chip interface temperature is therefore important in identifying cutting speeds where the built-up edge is minimum [4].

Diffusion Wear

When the temperatures of the tool and work materials increase at the contact zones, the atoms in the two materials become restive and migrate to the opposite material where the concentration of the same atom is less. Typically, in a tool material such as tungsten carbide (WC), where carbide (C) provides the hardness, while cobalt (Co) binds the WC grains, carbon diffuses to the moving steel chips, which have less concentration of the same atoms. Progressive diffusion of tool materials into the chip gradually leads to a weakened cutting edge and eventual chipping or breakage of the tool.

Oxidation Wear

The atoms in the cutting tool and/or work material form new molecules at the contact boundary where the area is exposed to the air (i.e., oxygen). Tungsten and cobalt in the cutting tool are oxidized close to the work surface–cutting tool flank, which leads to a notch wear on the cutting tool. Depending on the tool–work materials, tool geometry, and cutting conditions, one wear mechanism may be dominant, but all of them may occur simultaneously but at different rates. The wear of the tool localizes at two regions where the tool is in contact with the work material. The chip moves over the rake face until it leaves the contact area, where crater wear occurs. The freshly cut surface contacts the flank face of the tool, where the flank wear is observed (see Fig. 2.28).

Crater Wear

Crater wear occurs at the tool–chip contact area where the tool is subject to a friction force of the moving chip under heavy loads and high temperatures (Fig. 2.29). At higher speeds (i.e., turning P20 mold steel at $v = 250$ m/min cutting speed), the temperature on the rake face of a carbide tool may reach over $1,000°C$. At these high temperatures, the atoms in the tool continuously diffuse to the moving chip. The temperature is greatest near the midpoint of the tool–chip contact length, where the greatest amount of crater wear occurs due to intensive diffusion. As the crater wear approaches the cutting edge, it weakens the wedge and causes chipping of the tool. Crater wear can be minimized by selecting a tool material that has the least affinity to the workpiece material in terms of diffusion. The use of lubricants also reduces the wear. The lubricant penetrates between the chip and the tool, reducing the friction force and thus the temperature (see Eq. 2.27). As a result, the activity of diffusion is reduced leading to less crater wear. The chemical affinity between the workpiece material (mostly iron Fe) and the tool material (mostly tungsten-carbide WC) can be reduced using Al_2O_3, TiN, and TiC coatings on the tools. The thickness of the coating layer is typically 3 to 5 μm. The coating materials have low friction coefficients and have strong chemical stability at high temperatures. They act as a temperature barrier between the tool's core carbide material and the moving chip. However, they can not be easily deposited on sharp edges of the tool, and they may crack under heavy interrupted cutting conditions. The use of coated tools may significantly increase productivity by reducing tool wear at high speeds. Care must be taken when selecting coating materials that must not have chemical affinity with the contents of the workpiece material at elevated temperatures.

Flank Wear

Flank wear is caused by friction between the flank face (primary clearance face) of the tool and the machined workpiece surface (Fig. 2.30). At the tool flank–workpiece surface contact area, tool particles adhere to the workpiece surface and are periodically sheared off. Adhesion of the tool and workpiece materials increases at higher temperatures. Abrasive wear occurs when hard inclusions of work material or escaped tool particles scratch the flank and workpiece surface as they move across the contact area. Although adhesive and abrasive wear mechanisms are predominant in flank wear, some diffusion wear also exists.

After selecting the lubricant, cutting tool material, and geometry to be used in machining a particular work material, the best engineering approach is to identify a machinability speed that corresponds to the critical temperature limits of both cutting tool and work materials. The temperature can be estimated approximately by using the orthogonal cutting mechanics expressions given in this chapter (see Eq. 2.32). Alternatively, more accurate analytical and experimental techniques can be used to predict the tool–work material interface temperature as presented by Oxley [4] and Trent [39]. The tool–work material interface temperature must be below the diffusion and melting limits of the materials used in the cutting tool; otherwise binding materials such as cobalt in the WC or TiN and Al_2O_3 in cubic boron nitride (CBN) tools may diffuse toward the chip leading to a chain of diffusion, adhesion, and abrasive wear mechanisms. It is also advantageous to have a temperature that corresponds to at least that of the recrystallization phase of the work material where the strain hardening diminishes or even to that where the material becomes almost like a liquid with a significantly reduced yield shear strength. The built-up edge is avoided, and cutting forces and hence the stresses in the cutting tool are reduced when the material softens at elevated temperature, provided that the diffusion limits of the material within the cutting tool are not exceeded.

A sample investigation carried out by Ren and Altintas is summarized here as an example. A P20 mold steel with 34 Rc hardness is turned with a medium grain CBN tool without any lubricant. The CBN cutting tool had a 0.1 mm wide $15°$ negative chamfer angle with binders of TiN and Al_2O_3. P20 compositions are: 0.33%C, 0.3%Si, 1.4%Mn, 1.8%Cr, 0.8%Ni, 0.2%Mo, 0.008%S. P20 is widely used for injection molding dies. CBN cutting tools are the hardest (87Rc) after diamond, and owing to their high thermal and abrasion resistance, and low friction coefficient, they are widely used in high speed machining of hardened tool and die steels. Using the analytical cutting mechanics models, the average temperature at the shear plane and tool–chip interface are estimated for a given speed and feed [40]. It is found that the most optimal cutting speed in turning P20 with the CBN tool used is about 500 m/min, which corresponds to an average temperature of 1,400°C on the tool–chip interface. The diffusion limit for the CBN binding material (Al_2O_3) is about 1,600°C, and the melting temperature of P20 steel is about 1,300°C. For ISO S10 carbide tools, the diffusion limit of cobalt binding is about 1,300°C, which requires a cutting speed of less than 300 m/min without lubricant. A series of orthogonal cutting tests are conducted to verify the predictions, and three sample CBN tool pictures are shown in Figure 2.29. The photographs are taken with a scanning electron microscope (SEM). The CBN tool experienced the least amount of wear at a cutting speed of 240 m/min, where the tool–chip interface temperature is predicted to be about 1,150°C. The tool still does not exhibit any wear at 400 m/min after cutting a length of 907 m of work material. However, the temperature is very close to the melting limit of P20 steel, and chemical wear such as oxidation of Fe in steel with CBN tool materials may occur. When the speed is increased to 800 m/min, which corresponds to about 1,600°C at the tool-chip interface, severe crater wear can be observed from the third tool after a shorter machining length of 780 m. At this speed, the bindings of the CBN tool diffuse to the chip. The temperature prediction study indicates that the most optimal cutting speed for this particular medium grained CBN tool is about 500 m/min.

The classical wear mechanisms explained above are applicable to continuous machining operations such as turning. The milling process is a more complex operation owing to its intermittent nature, and thus additional factors affect tool wear in these operations. In milling, the cutting forces periodically change as the chip thickness varies. The tooth periodically enters and exits from the workpiece; hence it experiences stress and temperature cycling during cutting. As the tool enters the workpiece, it is heated. The tool starts cooling when it exits from the workpiece to the cutting environment (i.e., air or lubricant shower). The cooling period continues until it reenters the workpiece. This periodic thermal cycle produces alternating compressive (heating cycle) and tensile (cooling) stresses on the tool that may exceed its strength. Even if the thermal stress amplitudes are not large enough to break the tool suddenly, the thermal stress cycling causes gradual fatigue failure and wear of the tool. All of the wear mechanisms are therefore related to the tool–workpiece materials, the cutting forces acting on the tool, and the temperature in the contact regions. The temperature is a direct function of the relative speed and friction force between the materials in contact. Higher speed results in more friction energy, which then increases the temperature at the rake face–chip and flank face–workpiece surface contact zones. Thus, the cutting speed has the strongest influence on the wear mechanism. As flank wear increases, the tool–workpiece contact area becomes larger and hence rubbing on the workpiece surface becomes stronger. This results in a poor surface finish as well as high friction forces and temperatures, which may eventually lead to tool breakage. Flank wear occurs at the expense of losing a portion of the sharp cutting edge; therefore the accuracy of the finished workpiece dimension suffers equal to this amount. In practice, it is more important to control flank wear than crater wear. However, the volume that can be worn away before the total destruction of the tool occurs is much greater for crater wear than for flank wear, if there is not a strong chemical affinity between the tool and workpiece materials.

The flank wear land is measured as the width of wear land (VB) on the primary clearance face (Fig. 2.28). A typical tool life curve is shown in Figure 2.31. The history of flank wear with machining time may be split into three regions. The very sharp edge of the tool is worn soon after the cutting starts. This is followed by a gradual, approximately linear tool wear development with increasing cutting time. After the wear land (VB) reaches a critical limit, the flank wear exponentially increases. The tool must be replaced before reaching the critical limit (VB_{lim}) to avoid catastrophic tool failure. The corresponding cutting time to VB_{lim} is called the *tool life* and was expressed first by Taylor [41] as a function of cutting conditions:

$$T_t = C_t V^{-p'} c^{-q'}, \tag{2.123}$$

where T_t [min] is the tool life, V [m/min] is the cutting speed, and c [mm/rev] is the feed rate. C_t, p', and q' are constants for a given tool–workpiece material pair and are identified from machinability tests.

Figure 2.31: Tool life (T) curves: Variation of flank wear land (VB) with time at different cutting speeds (V).

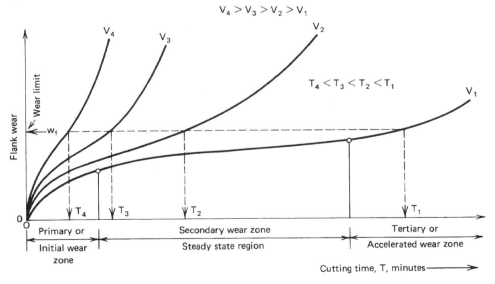

Example. The following measurements are obtained from the machinability tests, which are carried out by turning normalized AISI-1045 steel using grade K21 carbide inserts.

Test No.	Cutting Speed V (m/min)	Feed Rate (mm/rev)	Measured Tool Life T_t (min)
1	100	0.2	80
2	200	0.2	10
3	200	0.1	40

From tests 1 and 2,

$$p' = \frac{\ln(T_{t1}/T_{t2})}{\ln(V_2/V_1)} = 3.$$

From tests 2 and 3,

$$q' = \frac{\ln(T_{t3}/T_{t2})}{\ln(c_2/c_3)} = 2.$$

Substituting p' and q' into Eq. (2.123) gives the third parameter as

$$C_t = 3.2 \cdot 10^6.$$

The resulting empirical Taylor tool life equation for this particular workpiece material–tool couple becomes

$$3.2 \cdot 10^6 = T_t V^3 c^2.$$

For constant feed rate (c) or constant speed (V), logarithmic tool life charts can be prepared (Eq. 2.123) that can be used for appropriate selection of feeds and speeds for a desired tool life performance. It can be observed from the example that the feed rate has far less influence than the cutting speed (i.e., $V \gg c \rightarrow q' > p'$). The ideal situation is to monitor the progress of tool wear online during machining, so that the tool is replaced only when the wear limit is reached. However, there is no reliable and practical in-process tool wear measurement system available at the present. The topic is the subject of ongoing intensive research efforts. It is a well-known fact that the cutting forces increase, especially in the feed direction, as the flank wear grows. Friction between the flank and finish surface produces additional forces to the shearing. These forces have greater influence in the direction normal to the flank surface (i.e., the feed direction). However, correlating the increase in cutting force with tool wear is still difficult, since the cutting forces may also increase due to changes in the workpiece geometry and material properties of the workpiece.

2.10.2 Tool Breakage

Tool fracture is defined as the loss of a major portion of the tool wedge, which terminates the total cutting ability of the tool. Chipping of the tool (i.e., the loss of small particles from the cutting edge of the tool) is undesirable but does not prevent cutting totally. Chipping does, however, increase the friction on both the rake and flank faces of the chipped tool (Fig. 2.28). If it remains undetected, chipping eventually leads to total breakage of the tool.

Metal cutting tools are made of brittle materials that can fail under excessive chip loading (i.e., large feed rate) and fatigue owing to cyclic mechanical and thermal stresses. Experiments carried out by Zorev [42] showed the profile of tangential and normal load distribution on the tool–chip contact zone (see Fig. 2.2). Zorev found that when the chip moves away from the cutting edge, it sticks to the rake face first; this is followed by a sliding motion until it separates from the tool. Thus the tangential load in the sticking zone is equal to the yield shear stress of the material. In the sliding zone, the friction coefficient is constant and equal to the coefficient of friction between the tool and workpiece material. In the analysis of simplified orthogonal cutting mechanics, loading is assumed to be linear using average friction coefficient (μ_a) between the chip and the tool's rake face. Loladze [43] used a photo-elasticity method to measure the stress distribution within the tool wedge during cutting. He showed that after the tool–chip contact ends, the compressive stress zone in the tool is replaced by tensile stresses (see Fig. 2.2). Changes in the cutting conditions, especially in the uncut chip thickness, vary the stress distributions. As the uncut chip thickness is increased, the tensile stress zone expands and its magnitude increases.

Therefore, if the feed rate is increased until the principal tensile stresses reach the fracture limit (i.e., ultimate tensile strength of the brittle cutting tool materials), the tool starts to crack. Both carbide and high speed steel (HSS) tools may plastically fail as well owing to high temperatures at high cutting speeds. The tool loses its strength, deforms plastically, and is sheared off by cutting forces at the elevated temperatures. The plastic failure of the tool is most common in machining low machinability, heat resistant metals such as the titanium and nickel alloys used in aerospace industry. The chips of these alloys do not easily dissipate the heat away from the tool contact zone, thus causing large temperature loading of the tools. The friction energy, or heat created at the tool–chip contact zone, is directly proportional to the cutting speed (see Eq. 2.27). Therefore, while a free machining steel can be cut with a carbide tool at a typical cutting speed of 200 m/min, titanium or nickel alloy can be machined only at about 60 m/min. HSS tools can not be used in machining these alloys. Most of the coating materials used on carbide tools employ TiC or TiN, which have chemical affinity to titanium alloyed workpieces. With TiC/Al_2O_3 coated carbide tools in turning nickel-based alloys, flank wear rate is reduced substantially because (1) TiC acts as a heat barrier between the chip and the substrate carbide tool and (2) Al_2O_3 reduces the friction coefficient. The subjects of the fundamental mechanics of cutting, machinability, and prediction of cutting forces for different operations are only briefly introduced here. Each subject requires a dedicated text for in-depth understanding of the physics as well as the methods employed in modeling. Readers are referred to

TABLE 2.3. Orthogonal Cutting Test Conditions and Measurements in Plunge Turning of P20 Mold Steel

Feed Rate c [mm/rev]	Tangential Force F_t [N]	Feed Force F_r [N]	Measured Chip Thickness h_c [mm]
0.02	350	290	0.050
0.03	480	350	0.058
0.04	590	400	0.074
0.05	690	440	0.083
0.06	790	480	0.102
0.07	890	505	0.116
0.08	980	540	0.131

the well-established texts authored by Oxley [4], Armarego [2], Shaw [3], Trent [39], and Boothroyd [44].

2.11 PROBLEMS

1. A set of orthogonal tests are conducted to identify the shear angle, average friction coefficient, and shear stress of P20 mold steel that has a hardness of 34Rc. The cutting conditions and measured forces and chip thicknesses are given in Table 2.3. The cutting tool was an S10 grade plunge turning tool with a zero rake angle. The width of cut (i.e., the width of disk) was $b = 5$ mm, and the cutting speed was $V = 240$ m/min. The properties of P20 steel are given as: specific coefficient $c_s = 460$ Nm/kgC, specific density $\rho = 7,800$ kg/m^3, thermal conductivity $c_t = 28.74$ [W/mC].

(a) Evaluate the cutting coefficients K_{tc} and K_{fc} [N/mm^2] and edge force constants K_{te} and K_{fe} [N/mm] by a linear regression of the measured forces.

(b) Evaluate the shear angle (ϕ_c), shear stress (τ_s), and average friction coefficient (β_a) for each test, and express them as an empirical function of uncut chip thickness (h) to form an orthogonal cutting database.

(c) Predict the cutting force coefficients (K_{tc}, K_{fc}) using empirically expressed τ_s, ϕ_c, and β_a, and compare them against the values identified from mechanistic linear regression of the forces.

(d) Evaluate the shear strain and strain rate for each test at the primary shear zone.

(e) Identify the average temperature at the primary shear zone and tool–chip interface.

(f) Using the orthogonal to oblique transformation, express the cutting constants for a helical end mill that has 30° and 5° of helix and normal rake angles, respectively.

2. A P20 mold steel shaft with a diameter of $D = 57$ mm is turned with an S10 grade carbide tool with a side rake angle of $\alpha_f = -5°$, a back rake angle of $\alpha_p = -5°$, and zero approach angle. The nose radius of the insert is $r = 0.8$ mm. The radial depth of cut, the feed rate, and the cutting speed are $a = 1$ mm, $c = 0.06$ mm/rev, and $V = 240$ m/min, respectively. The orthogonal dry cutting parameters of P20 mold steel cut by an S10 carbide tool with a rake angle $\alpha_r = -5°$ are given as follows:

Shear stress $\qquad \tau_s = 1400h + 0.327V + 507$ [N/mm²],
Friction coefficient $\quad \beta_a = 33.69 - 12.16h - 0.0022V$ [degree],
Chip ratio $\qquad\qquad r_c = 2.71h + 0.00045V + 0.227$,

where the units are in h [mm] and V [m/min].

(a) Evaluate the distribution of chip thickness along the curved chip length.

(b) Evaluate the cutting coefficients along the curved chip length using the oblique cutting transformation.

(c) Evaluate the distribution of tangential, radial, and feed cutting forces (F_t, F_r, F_f) along the curved chip.

(d) Find the total cutting forces (F_x, F_y, F_z), torque, and power required to turn the shaft.

3. An end mill with 100 mm diameter, 4 flutes, 30° helix angle, and 10° rake angles is used in the peripheral milling of aerospace wing support frames. The material is Al-7075. A series of full immersion (i.e., slot) milling experiments are conducted at constant speed (2,500 rev/min) and axial depth of cut (a = 1.5 mm), but at a series of feed rates. The cutting forces are measured and the average forces per tooth period are given in Table 2.4. Calculate the cutting constants (K_{tc}, K_{rc}, K_{ac}) and edge constants (K_{te}, K_{re}, K_{ae}) by assuming the force model given in Eq. (2.79).

4. Write a simulation program for general milling operations. The program must be flexible to test various face and end milling operations. Use the cutting constants identified from the previous question given above. Neglect the influence of helix angle on cutting constants. Your simulation

TABLE 2.4. Measured Average Milling Forces for Al7075

Feed Rate [mm/th]	$\overline{F}_x$ [N]	$\overline{F}_y$ [N]	$\overline{F}_z$ [N]
0.025	−69.4665	64.3759	−10.2896
0.050	−88.1012	99.1383	−20.5814
0.100	−105.6237	155.7100	−33.9852
0.150	−118.1107	210.4644	53.4833
0.200	−130.6807	263.4002	73.1515

TABLE 2.5. Simulation Cases

Case No.	a (mm)	β (deg)	ϕ_{st} (deg)	ϕ_{ex} (deg)
i	5.00	0	0	120
ii	40.0	30	0	90
iii	40.0	30	0	180

program must be able to cover at least one full revolution of the cutter. Show the graphical results for the milling cases given in Table 2.5

Feed rate $(c) = 0.1$ mm/tooth,   Spindle speed $(n) = 6{,}000$ rev/min,
Gauge length $(l) = 54.5$ mm,   Cutter diameter $(d) = 19.05$ mm,
No. of flutes $(N) = 4$,   Young's modulus $(E) = 2 \cdot 10^5$ MPa.

Plot the following variables on the same graph and provide short comments for the simulation results:

(a) F_x and F_y cutting forces, all cases.
(b) Resultant cutting force and torque, all cases.

5. A twist drill with a diameter $2R = 19.05$ mm, chisel edge angle $\psi_c = 125°$, chisel width $2w = 0.78$ mm, and nominal helix angle of $\beta_0 = 30°$ is used in drilling titanium alloy Ti_6Al_4V. Predict the thrust, torque, and power of the drilling operation when the drill feed and spindle speed are $c = 0.1$ mm/rev and $n = 200$ rev/min, respectively. Use the orthogonal cutting parameters of Ti_6Al_4V given in Table 2.1.

6. A CBN tool is used in dry turning of P20 mold steel with 34Rc hardness. The cutting edge has a 15° chamfer edge, which acts like a negative rake angle. A number of turning tests are conducted at different feeds and speeds. Each test is continued until a cutting length of l_m is reached, and the average flank wear (VB) is recorded. Using a least squares method, the following relationship was identified from an actual cutting test data:

$$VB = 0.0023568 V^{1.33} c^{0.7423} l_m^{0.796},$$

where the units are: flank wear VB [mm], cutting speed V [m/min], feed rate c [mm/rev], and machining length l_m [m]. Predict the tool change time when the cutting conditions are selected as: $V = 240$ m/min and $c = 0.06$ mm/rev and the maximum allowable wear is $VB = 0.2$ mm. The diameter of the shaft is 100 mm.

CHAPTER THREE

STATIC AND DYNAMIC DEFORMATIONS IN MACHINING

3.1 INTRODUCTION

Machine tools are called *machine making machines*. Various machining and forming operations are executed by a variety of machine tools to produce mechanical parts. In order to maintain specified tolerances, the machine tools must have greater accuracy than the tolerances of the manufactured parts. The precision of a machine tool is affected by the positioning accuracy of the cutting tool with respect to the workpiece and the relative structural deformations between them. The engineering analysis and modeling of relative static and dynamic deformations between the cutting tool and workpiece are covered in this chapter.

3.2 MACHINE TOOL STRUCTURES

A machine tool system has three main groups of parts: mechanical structures, drives, and controls. The components can be observed from the horizontal CNC machining center shown in Figure 3.1.

Mechanical Structure. The structure consists of stationary and moving bodies. The stationary bodies include beds, columns, bridges, and gear box housings. They usually carry moving bodies, such as tables, slides, spindles, gears, bearings, and carriages. The structural design of machine tool parts requires high rigidity, thermal stability, and damping. In general, the dimensions of machine tools are overestimated in order to minimize static and dynamic deformations during machining. The general design of machine tool structures will not be covered in this text. Instead, it is assumed that the relative static and dynamic compliance between the tool and the workpiece is measured experimentally or predicted with analytical methods. The effect of structural compliance on the accuracy of the machined workpiece and machining performance is presented.

Drives. Moving mechanisms are grouped into spindle and feed drives in machine tools. The spindle drive provides sufficient angular speed, torque, and power to a rotating spindle shaft, which is held in the spindle housing with roller or magnetic bearings. Low- to medium-speed spindle shafts are connected to the electric motor via V belts. There may be a single-step gear reducer and a clutch between the electric motor and spindle shaft. High-speed spindles

Figure 3.1: A horizontal CNC machining center. Source: Makino Milling Machine Co., Ltd.

($n > 15,000$ rev/min) may have electric motors that are built into the spindle to reduce the inertia and friction produced by the motor–spindle shaft coupling. In a typical and versatile machining center, the high- and low-speed spindles can be swapped in a short time period (i.e., one hour). The feed drives carry the table or the carriage. In general, the table is connected to the nut, and the nut houses a lead screw. The screw is connected to the drive motor either directly or via a gear system depending on the feed speed, inertia, and torque reduction requirements.Conventional machine tools have numerous gear reduction steps in order to obtain a desired feed speed. Each feed screw has a dedicated drive motor in CNC machine tools. Very high speed machine tools may employ linear direct motors and drives without the feed screw and nut, thus avoiding excessive inertia and friction contact elements.

Controls. The control parts include motors, amplifiers, switches, and computers that are used to energize the electrical parts in a controlled sequence and time. Conventional machine tools mostly contain relays, limit switches, and operator controlled potentiometers and directional control switches. CNC machine tools have power servo amplifiers, opto-isolated ON/OFF and limit switches, and a computer unit equipped with emergency, control, and operator interface units. The feed velocity and positioning accuracy of feed drives are dependent on the torque and power delivery of the servomotors and on the feed drive servocontrol algorithms executed in the computer numerical control (CNC) unit of the machine tool, which is covered in Chapters 4 and 5.

3.3 DIMENSIONAL FORM ERRORS IN MACHINING

The source of relative deformations between the cutting tool and the workpiece at the contact point may be due to thermal, weight, and cutting loads. The machine tool has rotating and moving elements, which are heated by friction energy. The temperature increase is never uniform among the machine elements, owing to both their varying thermal coefficients and the location of heat sources. The machine tool table spindle head, on an overhang portal frame or on a vertical column, changes location as the coordinate of the cutting point varies on the workpiece. The moving weights change the stiffness or the relative displacement between the tool and workpiece at the cutting point. Thermal and weight deformations can be measured at various positions of the machine tool table or spindle head and stored in the compensation registers of the CNC unit. When the machine tool travels, the errors are read from the registers and additional motions are ordered by the CNC to compensate the weight and thermal deformation errors during machining.

The deformation errors caused only by the cutting forces will be presented in this text. As the cutting tool travels along a tool path, there may be variations in the magnitude and the direction of cutting forces as well as the relative stiffness between the cutting tool and the workpiece [45]. The relative displacements cause deviations from the desired dimensions of the workpiece, causing *dimensional form errors*.

3.3.1 Form Errors in Cylindrical Turning

A typical cylindrical turning operation is shown in Figure 3.2. If the cutting tool has a nonzero approach angle or a significant nose radius, the turning operation has an oblique geometry where the radial force perpendicular to the finished surface is not zero. Any flexibility in the direction of radial force produces a relative displacement between the workpiece and the cutting tool at a point where the finish surface is generated. The structure can be modeled as an elastic circular shaft supported by a pin at the tail stock center and a pin support and resistive bending moment in the chuck at the spindle side. For simplicity, an average diameter (d) of the shaft can be used to account for the radial depth of cut removed. If the radial cutting force F_r is applied l distance away from the chuck (see Fig. 3.2), the radial deflection of the beam at an axial location x is

$$y(x) = \frac{1}{EI}\left[-\frac{R_c}{6}x^3 + \frac{F_r}{6}\langle x - l \rangle^3 + \frac{M_c}{2}x^2 \right] \tag{3.1}$$

where E is the modulus of elasticity and $I = (\pi d^4)/64$ is the moment of inertia of the workpiece. The support force R_c and bending moment M_c at the chuck

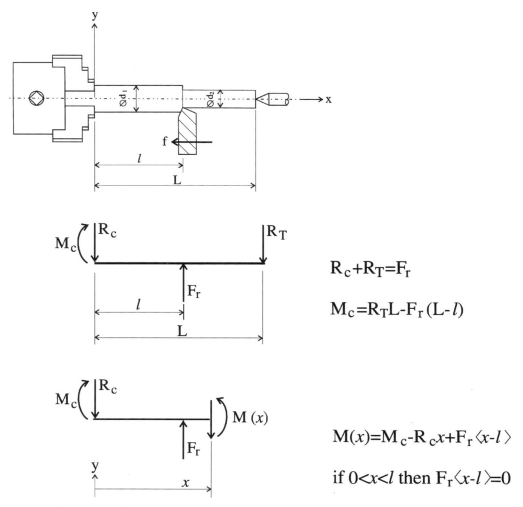

Figure 3.2: Form errors in cylindrical turning of a shaft.

are given by

$$R_c = \frac{F_r(L-l)}{2L^3}[3L^2 - (L-l)^2], \qquad M_c = \frac{F_r l(L-l)(2L-l)}{2L^2}.$$

The deflection, which is passed on the turned shaft as a dimensional error, occurs at the point where the tool is in contact with the workpiece (i.e., $x = l$). The radial deflection $y(l)$ is defined from the axis drawn between the center of the tail stock and that of the spindle (i.e., machine tool axis). Since the bar stock is pushed away by amount $y(l)$ in the radial direction from the stationary tool, the actual radial depth of cut will be $a - y(l)$. Hence the radius of the finished workpiece will have $y(l)$[mm] amount of extra stock or *form error* at location $x = l$. Note that as the tool travels toward the chuck, the stiffness of the beam varies and hence the form error left on the finished surface will be different along the

workpiece axis. The maximum form error occurs at the middle of the workpiece ($x = L/2$).
Note that the finished workpiece will not be
cylindrical any more but will have a barrel
shape. When the barrel type workpiece is
turned again in the subsequent pass, the
effective axial depth of cut will no longer be
uniform but will vary along the bar axis i.e.,
$a_{\text{effective}} = a + y(l)$. Since the amplitude of the
cutting force is proportional to the axial depth

Figure 3.3: Bore form errors produced by a boring bar.

of cut, the radial cutting force and the stiffness of the workpiece will vary along
the bar axis, thus still leaving a form error. It is therefore recommended to have
a small depth of cut and feed rate (i.e., small force) in the finishing operations.
The process planning engineer should calculate the feed and axial depth of cut
according to the tolerance and stiffness in the center of the workpiece.

3.3.2 Boring Bar

A general diagram of a boring operation is shown in Figure 3.3. The boring
bar resembles a cantilevered elastic beam, and it is held in a variety of configurations depending on the machine tool and workpiece set up. The boring bar
is held in the stationary tail stock of a lathe, and the workpiece is attached to
the rotating spindle. An enlarging internal diameter of a cylindrical shaft is a
typical example for a boring operation on a lathe. The boring bar can be held
on the spindle of a horizontal boring or machining center, but it is offset by an
amount equal to the radius of the bore to be opened. The offset is adjusted with
an *eccentric* attached to the spindle. Large bores of prismatic parts are opened
on the boring centers by placing the workpiece on the machine tool table [46].

If the length of the boring bar is l, the deflection at the workpiece–cutting
tool contact point will be

$$\delta = \frac{F_r l^3}{3EI} = \frac{F_r}{k_r}, \tag{3.2}$$

where F_r is the radial force and $k_r = (3EI)/l^3$ is the radial stiffness of the
boring bar. The radial deflection is passed as a form error to the bore, and
its magnitude depends on the amplitude of the cutting force and the stiffness
in the radial direction. If a cylindrical boring bar is used, $I = \pi d^4/64$ and
$k_r = (3\pi E d^4)/(64 l^3)$. If a boring bar has a rectangular section with a width of
b and a height of h, the moment of inertia is $I = (bh^3)/12$. The stiffness of the
rectangular boring bar becomes

$$k_r = \frac{Ebh^3}{4l^3}.$$

To increase the radial stiffness, the cutting tool must be mounted on the face of
boring bar section that has a width of h. If the ratio is $h/b > 1$, the magnitude
of the form error left on the surface is reduced by h^3 times by simply orienting
the rectangular bar in the stiffer direction.

3.3.3 Form Errors in End Milling

The walls of parts are produced by the periphery of end mills and thus end milling is also called *peripheral milling*. The finished workpiece surface is perpendicular to the direction of feed in end milling. If the feed and normal directions are aligned with the Cartesian x and y axes respectively, any deflection in the y axis may produce a static form error. End mills can be considered as an elastic cylindrical beam, cantilevered to the spindle through collet and chuck. They are generally the most flexible part in the machine tool system, since their aspect ratio, the ratio of diameter over the gage length from the collet, is rather small.

The form errors produced by helical end mills are rather complex [47]. The cutting forces are not constant but vary with the rotation of the end mill. Furthermore, the helix angle of the flutes produces additional variation on the distribution of cutting forces along the cutter z axis [48].

First consider an end mill with straight flutes (i.e., zero helix angle) for simplicity in explaining the surface generation mechanism. Here the deflections perpendicular to the finish surface (in the normal direction y) are important. The static deflection caused by a normal force (F_y) at the free end of the end mill is given by [49]

$$\delta_y = \frac{F_y}{k}, \tag{3.3}$$

where $k = (3EI)/l^3$ and $I = (\pi d^4)/64$. The effective diameter of the cutter is d, and the gage distance is l from the collet. The effective diameter of the cutter can be found by scaling its outer diameter by 0.8–0.85 to take flutes into account. The cutting force is proportional to the chip thickness h, that is,

$$F_y(\phi) = K_t a h(\phi)$$
$$\times [\cos(\phi) + K_r \sin(\phi)],$$

where ϕ is the immersion angle measured from the y axis, the chip thickness is $h(\phi) = c \sin\phi$, and c is the feed per tooth. The tooth generates the surface when it is in contact with it, or when it is on the normal y axis. When the cutting tool lies on the y axis, the chip thickness is always zero. This situation occurs when the tool is at the entry in up-milling ($\phi = 0$) or at the exit in down-milling ($\phi = \pi$) (see Fig. 3.4). Thus, when

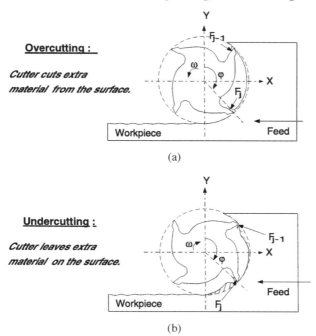

Figure 3.4: The influence of the width of cut and milling mode on the surface form errors: (a) up milling; (b) down milling.

only one tooth is present in the cutting zone, regardless of the flexibility of the end mill or magnitude of the cut, the surface form errors will be zero in end milling with straight flutes. This is a common situation in finishing cuts where the radial immersion angle is usually much less than the cutter pitch angle. However, if there are two or more teeth cutting simultaneously, the cutting force will not be zero when one of the teeth is aligned with the y axis, since the other flutes cut a chip within the immersion zone. In this case, the tooth either deflects toward the surface in up-milling, thus causing an *overcut form error*, or deflects away from the surface in down-milling, causing an *undercut form error*.

Generating the surface becomes complex when the end mill has helical flutes. Even when there is only one flute in cut, there will always be form errors left on the surface [50]. Consider a flute's edge at the bottom of the cutter aligned with the normal y axis, thus having a zero immersion (i.e., $\phi(z=0)=0$). As the end mill rotates, the tip of the flute moves to immersion ϕ, while an upper point on the flute, which has a z coordinate measured from the tip, will be right on the y axis, hence generating the surface. Since the normal cutting force will not be zero at this instant, the elastic end mill displacement will produce a form error on the surface. As the cutter rotates, the cutting edge point that generates the surface will move upward along the flute because of the helix angle. Depending on the number of flutes and width of cut, there may be more than one cutting edge point aligned at the y axis or in contact with the finish surface. The contact points can be calculated by equating the instantaneous immersion angle ($\phi_j(z) = \phi + (j-1)\phi_p - k_\beta z$, with $k_\beta = (2\tan\beta)/d$) to zero in up-milling and to π in down-milling [32]:

$$z = \frac{\phi + (j-1)\phi_p}{k_\beta} \quad \text{(up-milling)},$$

$$z = \frac{\pi - (\phi + (j-1)\phi_p)}{k_\beta} \quad \text{(down-milling)},$$

where β is the helix angle, $j = 1, 2, \ldots$ $N-1$ is the flute index, and $\phi_p = (2\pi)/N$ is the cutter pitch angle. An algorithm that can predict the surface form errors can be integrated to the cutting force prediction program presented in the previous chapter. The cutter can be divided into M number of small disk elements within the axial depth of cut a (see Fig. 3.5), and it can be rotated at increments $\Delta\phi$, (i.e., $\phi = 0, \Delta\phi, 2\Delta\phi, \ldots, \phi_p$). Each differential element has an axial depth of cut $\Delta z = a/M$, and the influence of the helix angle may be neglected by selecting

Figure 3.5: Static deformation model of an end mill.

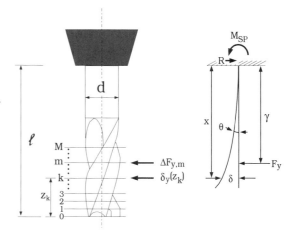

small elements. The differential cutting force produced by element m is given by (Fig. 3.5)

$$\Delta F_{y,m}(\phi) = K_t c \Delta z \sum_{j=0}^{N-1} [\sin \phi_j(z) - K_r \cos \phi_j(z)] \sin \phi_j, \qquad (3.4)$$

where K_t and K_r are cutting constants and c is the feed rate per tooth.

The immersion angle for the element m is $\phi_j(m) = \phi + (j-1)\phi_p - k_\beta \cdot m\Delta z$. The elemental cutting forces may be lumped at the upper boundaries of disks. The deflection in the y direction at the contact point z_k caused by the force applied at element m is given by the cantilevered beam formulation as [51]

$$\delta_y(z_k, m) = \begin{cases} \dfrac{\Delta F_{y,m} v_m^2}{6EI}(3v_m - v_k), & 0 < v_k < v_m, \\[2ex] \dfrac{\Delta F_{y,m} v_m^2}{6EI}(3v_k - v_m), & v_m < v_k, \end{cases} \qquad (3.5)$$

where E is Young's Modulus, I is the area moment of inertia of the tool, and $v_k = l - z_k$, with l being the gage length of the cutter measured from the collet face. The area moment of the tool is calculated using an equivalent tool radius of $R_e = 0.80R$, where 0.8 is the approximate scale factor due to flutes. The total static deflection at axial contact point z_k is calculated by the superposition of the deflections produced by all M elemental forces on the end mill:

$$\delta_y(z_k) = \sum_{m=1}^{M} \delta_y(z_k, m). \qquad (3.6)$$

At the points where the cutting edge is in contact with the finish surface, the deflection $\delta_y(z_k)$ is imprinted as a dimensional error on the workpiece. By rotating the cutter, the dimensional errors along the finish surface wall can be generated and displayed on the plane of the cutter axis and surface normal (y, z). The three-dimensional topography of the workpiece is generated by simply repositioning the cutter along the feed direction (x). Figure 3.6 shows a predicted surface error map of a workpiece that has a varying width of cut along the feed direction. A detailed analysis of dimensional errors in end milling can be found in Reference [51].

3.4 STRUCTURAL VIBRATIONS IN MACHINING

Machine tool vibrations play an important role in hindering productivity during machining. Excessive vibrations accelerate tool wear and chipping, cause poor surface finish, and may damage the spindle bearings [52]. A brief review of basic vibration theory is provided first. As experimental modal analysis techniques are most readily used in modern machining facilities, a practical review of modal analysis theory and its practice in machining is then presented. The basics of vibration engineering should enable the reader to understand machine tool vibrations and their avoidance in practice.

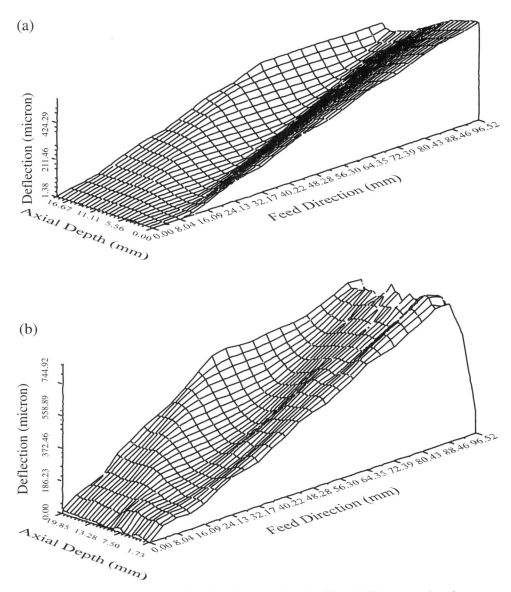

Figure 3.6: A sample view of predicted surface area in end milling. (a) Dimensional surface error simulation results; (b) experimental results.

3.4.1 Fundamentals of Free and Forced Vibrations

A simple structure with a single degree of freedom (SDOF) system can be modeled by a combination of mass (m), spring (k), and damping (c) elements as shown in Figure 3.7. When an external force $F(t)$ is exerted on the structure, its motion is described by the following differential equation:

$$m\ddot{x} + c\dot{x} + kx = F(t). \tag{3.7}$$

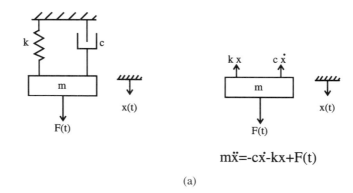

$$m\ddot{x} = -c\dot{x} - kx + F(t)$$

(a)

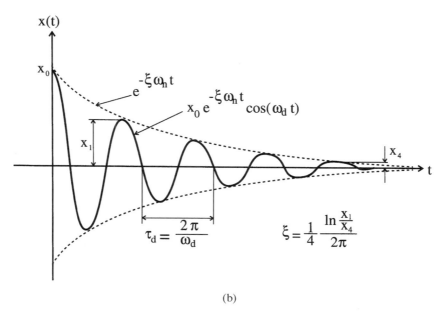

(b)

Figure 3.7: Mass, spring, and damping model of a single degree of freedom system: (a) SDOF system; (b) free vibration of an SDOF system.

If the system receives a hammer blow for a very short duration, or when it is at rest and statically deviates from its equilibrium and lets go, the system experiences free vibrations. The amplitude of vibrations decays with time as a function of the system's damping constant. The frequency of the vibrations are mainly dominated by the stiffness and the mass and are lightly influenced by the viscous damping constant, which is very small in mechanical structures. When the damping constant is zero ($c = 0$), the system oscillates at its natural frequency

$$\omega_n = \sqrt{\frac{k}{m}}.$$

A damping ratio is defined as $\zeta = c/2\sqrt{km}$, which is always less than one in mechanical structures. In most metal structures $\zeta < 0.05$ or even less. The damped natural frequency of the structure is defined by

$$\omega_d = \omega_n \sqrt{1 - \zeta^2}.$$

Assuming that the mass is *free* from any external force and displaced statically by an amount of x_0 and that the system is released, the free vibration motion can be described by

$$x(t) = x_0 e^{-\zeta \omega_n t} \cos \omega_d t.$$

The period between each vibration wave is $\tau_d = 2\pi/\omega_d$, which is a simple way of estimating the system's damped natural frequency from the *free* or *transient* vibration measurements. The damping constant can be approximated from the ratio of decaying maximum amplitudes of first and nth successive waves with the equation

$$\zeta = \frac{1}{n}\left(\ln \frac{x_1}{x_n}\right)\Big/ 2\pi.$$

When an external force $F(t)$ is present, the system experiences *forced vibrations*. When a constant force $F(t) = F_0$ is applied to the structure, the system experiences a short-lived *free* or *transient* vibration and then stabilizes at a static deflection $x_{st} = F_0/k$. Assuming that the external force is harmonic (i.e., can be represented by sine or cosine functions or their combinations), we can write

$$m\ddot{x} + c\dot{x} + kx = F_0 \sin(\omega t). \tag{3.8}$$

Thus the system experiences *forced* vibrations at the same frequency ω of the external force, but with a time or *phase delay*. Let us assume that transient vibrations caused by initial loading have diminished and the system is at *steady-state* operation. Then

$$x(t) = X \sin(\omega t + \phi).$$

It is mathematically more convenient to use complex harmonic functions in forced vibrations. The harmonic force can be expressed by $F(t) = F_0 e^{j\alpha} e^{j\omega t}$, where α is the phase lead from a reference time or angular position on a complex plane. The corresponding harmonic response is $x(t) = X e^{j(\omega t + \phi)}$, and when this is substituted into the equation of motion (3.8) we get

$$(k - \omega^2 m + j\omega c)X e^{j\phi} e^{j\omega t} = F(t) = F_0 e^{j\alpha} e^{j\omega t}.$$

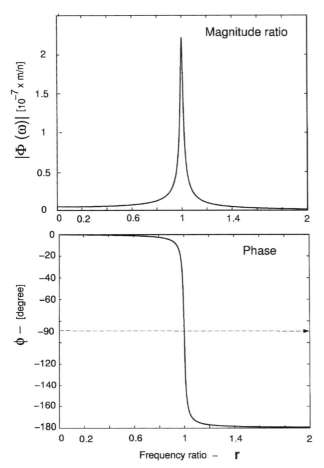

Figure 3.8: Transfer function of a SDOF system.

The resulting amplitude and phase of the harmonic vibrations are

$$|\Phi(\omega)| = \left| \frac{X}{F_0} \right|$$

$$= \frac{1}{k} \frac{1}{\sqrt{(1 - r^2)^2 + (2\zeta r)^2}},$$

$$\phi = \tan^{-1} \frac{-2\zeta r}{1 - r^2} + \alpha,$$

$$(3.9)$$

respectively, where the excitation to natural frequency ratio is $r = \omega/\omega_n$. Equation (3.9) is called the *transfer function, frequency response function*, or *receptance* of the SDOF structure, and its graphical illustration is shown in Figure 3.8. The transfer function ($\Phi(\omega)$) can be separated into real ($G(\omega)$) and imaginary ($H(\omega)$) components of $\frac{X}{F_0} e^{j(\phi - \alpha)}$ as follows:

$$G(\omega) = \frac{1 - r^2}{k[(1 - r^2)^2 + (2\zeta r)^2]},$$

$$H(\omega) = \frac{-2\zeta r}{k[(1 - r^2)^2 + (2\zeta r)^2]},$$

$$(3.10)$$

and $\Phi(\omega) = G(\omega) + jH(\omega)$. Note that at resonance ($\omega = \omega_n$, $r = 1$), $G(\omega_n) = 0$, $H(\omega_n) = -1/(2\,\text{kg})$.

The real and imaginary parts of the transfer function are illustrated separately in Figure 3.9, and by a polar plot in Figure 3.10. At zero frequency crossing, the real part is equal to the static flexibility ($1/k$). As the excitation frequency approaches the natural frequency (i.e., $r = 1$), the system resonates, the amplitude of the vibrations become maximum, and the phase angle approaches -90 degrees. The time delay between the excitation and response can be evaluated by $t_d = \phi/\omega$ at the harmonic excitation frequency ω. If the excitation frequency continues to increase, the phase angle approaches -180 degrees, or the delay becomes half a period of excitation. The amplitude of the vibrations decreases, because the physical structure can not respond to high-frequency disturbances. The damping ratio, stiffness, and natural frequency can be estimated from the transfer function measured with a Fourier analyzer. At zero excitation frequency ($\omega = 0$), the magnitude of $\Phi(\omega)$, and the real receptance $G(\omega)$ are equal to the static flexibility ($1/k$). Care must be taken in reading this value at low frequency because of the poor sensitivity of measurements taken with velocity and acceleration sensors. An extrapolation of the transfer function from higher frequencies where the resonance levels off can be considered as an alternative method to estimate the stiffness. Displacement sensors provide more accurate measurement of static flexibility. The maximum magnitude of $\Phi(\omega)$ occurs at $\omega = \omega_n \sqrt{1 - 2\zeta^2}$. The real receptance $G(\omega)$ has two extrema at

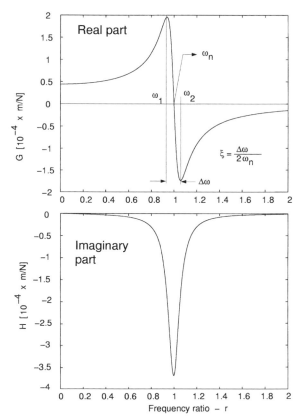

Figure 3.9: Transfer function of a SDOF system represented by its real and imaginary parts.

$$\omega_1 = \omega_n \sqrt{1 - 2\zeta} \rightarrow G_{max} = \frac{1}{4k\zeta(1 - \zeta)},$$

$$\omega_2 = \omega_n \sqrt{1 + 2\zeta} \rightarrow G_{min} = -\frac{1}{4k\zeta(1 + \zeta)}. \tag{3.11}$$

In practical machinery, the external excitations are usually periodic but not harmonic. Any periodic force can be represented by its harmonic components. When the external force $F(t)$ (i.e., milling force) is periodic with a period of $\tau = 2\pi/\omega$ (i.e., tooth period), it can be expanded in a Fourier series:

$$F(t) = \frac{a_0}{2} + \sum_{n=1}^{\infty} a_n \cos n\omega t + \sum_{n=1}^{\infty} b_n \sin n\omega t, \tag{3.12}$$

where n is the harmonic of the fundamental frequency ω. The exact solution of Fourier coefficients can be found by taking continuous integrals, which require a mathematical representation of the periodic forcing function $F(t)$. Because practical external excitation, such as milling forces, have periodic but irregular wave forms, a discrete numerical technique is used to calculate the Fourier coefficients. Assuming that the periodic excitation is uniformly digitized at

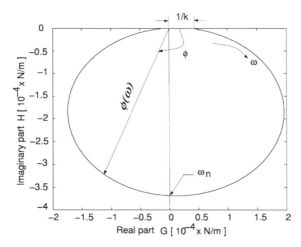

Figure 3.10: Transfer function of a SDOF system represented in polar coordinates.

T [s] intervals for N times per period τ (i.e., $\tau = NT$) we have

$$a_0 = \frac{2}{N} \sum_{i=1}^{N} F_i,$$

$$a_n = \frac{2}{N} \sum_{i=1}^{N} F_i \cos \frac{n 2 \pi t_i}{\tau},$$

$$n = 1, 2, 3, \ldots,$$

$$b_n = \frac{2}{N} \sum_{i=1}^{N} F_i \sin \frac{n 2 \pi t_i}{\tau},$$

$$n = 1, 2, 3, \ldots.$$

In experimental analysis, F_i corresponds to the ith sample of the force by the Fourier analyzer. An alternative discrete Fourier series representation of the periodic function is

$$F(t) = \sum_{n=0}^{N} c_n e^{-j \alpha_n} e^{j n \omega t}, \tag{3.13}$$

where

$$c_n = \sqrt{a_n^2 + b_n^2}, \qquad \alpha_n = \tan^{-1} \frac{b_n}{a_n}.$$

The plot of c_n, α_n versus frequency is called a *Fourier spectrum*. For a periodic but nonharmonic excitation, the steady-state response of an SDOF system can be calculated by superposing the vibration produced by each harmonic component of the periodic excitation:

$$x(t) = \sum_{n=0}^{N} \frac{c_n}{k \sqrt{(1 - n^2 r^2)^2 + (2 \zeta n r)^2}} e^{j(n \omega t - \alpha_n - \phi_n)}. \tag{3.14}$$

In most practical cases, the external periodic excitation can be approximated by only its first four to five harmonic (n) components, and usually the higher harmonics may not have sufficient energy to cause considerable influence on the vibrations. Milling forces, which are periodic at tooth passing frequency, can be represented by Fourier series components. If a harmonic of the milling force is close to one of the natural modes of the structure, an alternative spindle speed may be selected to avoid forced vibrations.

3.4.2   Oriented Transfer Function

The static and dynamic deformations between the tool and cut surface determine the accuracy and reliability of the manufactured component. A typical machine tool can be described by a series of masses interconnected by springs in different directions [45]. The resultant cutting force is transmitted to the machine via the springs and masses. The superposition of the displacements produced by all springs in the direction perpendicular to the cut surface determine the resulting dimensional accuracy and chip volume removed from the machined part.

Consider a common mass (m) connected to the rigid ground by a series of springs as shown in Figure 3.11. Each spring (i) and the mass define an SDOF system in each independent direction indicated by its angular orientation (θ_i) from the cut surface direction y. The resultant cutting force F has angle β with the cut surface direction y. The displacement produced in direction y by all springs that feel the force F is required.

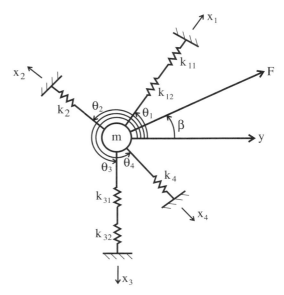

Figure 3.11: Cross transfer function of oriented displacement (y) and force (F).

The force transmitted to spring i is

$$F_i = F \cos(\theta_i - \beta).$$

If the transfer function in each spring direction is $\Phi_i(\omega) = x_i / F_i$, the corresponding displacement of spring i is

$$x_i = F_i \Phi(\omega) = F \cos(\theta_i - \beta)\Phi_i(\omega).$$

The displacement in direction y produced by spring i is

$$y_i = x_i \cos \theta_i = F \cos \theta_i \cos(\theta_i - \beta)\Phi(\omega).$$

Superposing the components of vibrations x_i in direction y gives the resultant vibrations. The transfer function between the cutting force F and the resultant vibration y, called the *cross* or *oriented* transfer function by Tlusty [45], is

$$\Phi_{yF}(\omega) = \frac{y}{F} = \sum_{i=1}^{4} u_{di} \Phi_i(\omega), \tag{3.15}$$

where $u_{di} = \cos \theta_i \cos(\theta_i - \beta)$ is the *directional factor* in direction i. When the mass is neglected, the preceeding formulation can be used to calculate the resultant static deformation in direction y.

3.4.3 Design and Measurement Coordinate Systems

There are three coordinate systems in the analysis of machine tool structures that are modeled by a multiple number of spring, damping, and mass elements.

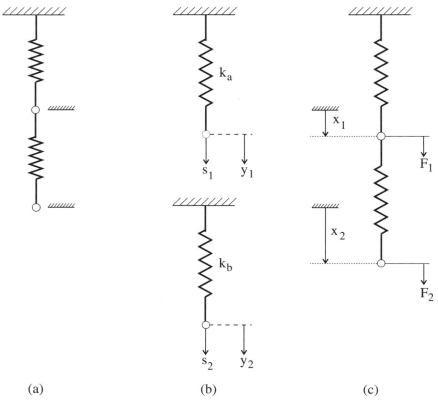

(a)   (b)   (c)

Figure 3.12: Displacements and forces defined in local and design coordinates: a) Connected springs; b) design coordinates; c) local coordinates.

These are *design, local,* and *modal* coordinate systems. The masses, spring, and damping constants may be defined in any of the three coordinate system, depending on the convenience of either computation or physical interpretation. The modal coordinates do not have any physical meaning. They are used to analyze the strength and behavior of the whole structure at a particular natural frequency, as is explained in this section.

   The design and local coordinates are explained using a simple structure represented by two springs connected in series (see Fig. 3.12). Assume that forces F_1 and F_2 are applied on the two connected springs, and the displacements are measured at fixed locations (i.e., at the initial points 1 and 2 where the springs were undeformed). The displacements are x_1 and x_2, and the reference points for the measurements are fixed. The displacements x_1 and x_2 are defined in *measurement* or *local coordinates*. However, when the relative elongation of each spring between its two ends is measured, the corresponding displacement is defined in *design coordinates*. In other words, while the local displacement represents the absolute change in the coordinate of a point on the structure, the design displacement represents elongation or compression of an individual spring [53].

Let the forces be S_1 and S_2 and the displacements be y_1 and y_2 in design coordinates. Thus

$$S_1 = k_a y_1, \qquad S_2 = k_b y_2,$$

or, in general matrix form,

$$\begin{Bmatrix} S_1 \\ S_2 \end{Bmatrix} = \begin{bmatrix} k_a & 0 \\ 0 & k_b \end{bmatrix} \begin{Bmatrix} y_1 \\ y_2 \end{Bmatrix},$$

where k_a and k_b are the design stiffness values for individual springs. For simplicity, we can use the following matrix notation:

$$\{S\} = [K_y]\{y\}, \tag{3.16}$$

where $\{S\}$ and $\{y\}$ are force and displacement vectors and $[K_y]$ is the stiffness matrix in the design coordinates. The relation between the design and local displacements can be expressed as

$$x_1 = y_1 \rightarrow y_1 = x_1,$$

$$x_2 = y_1 + y_2 \rightarrow y_2 = -x_1 + x_2,$$

or

$$\begin{Bmatrix} y_1 \\ y_2 \end{Bmatrix} = \begin{bmatrix} 1 & 0 \\ -1 & 1 \end{bmatrix} \begin{Bmatrix} x_1 \\ x_2 \end{Bmatrix} \rightarrow \{y\} = [T]\{x\}, \tag{3.17}$$

where matrix $[T]$ is the transformation matrix between the *local* and *design* displacements. A similar expression can be written for the forces, that is,

$$S_2 = F_2 \rightarrow F_2 = S_2,$$

$$S_1 = F_1 + F_2 \rightarrow F_1 = S_1 - S_2,$$

or

$$\{F\} = [T]^T\{S\}. \tag{3.18}$$

The stiffness, mass, and damping elements can be transformed from one coordinate system to another using the transformation matrix $[T]$. Substituting Eqs. (3.16) and (3.17) into (3.18) yields

$$\{F\} = [T]^T\{S\} = [T]^T[K_y]\{y\} = [T]^T[K_y][T]\{x\}.$$

Noting that $\{F\} = [K_x]\{x\}$, we can obtain the local stiffness matrix from the design coordinates as

$$[K_x] = [T]^T[K_y][T]. \tag{3.19}$$

Similar transformations can be applied to the damping and mass matrices, that is, $[C_x] = [T]^T[C_y][T]$, $[M_x] = [T]^T[M_y][T]$.

Note that depending on the displacements and forces, the units may be different in the design and local coordinates [53]. For example, the displacements and forces may be defined in mm and in local coordinates at specific locations of a structure, whereas the displacements may be angular if produced by torsional loads and springs in the design coordinates. Although all the experimental measurements are done in local coordinates, the design engineer is interested in determining individual weak elements defined in the design co-ordinates.

3.4.4 Analytical Modal Analysis for Multidegree-of-Freedom Systems

Machine tools have multiple degrees of freedom (DOF) in various directions. The vibrations between the cutting tool and generated workpiece surface are the main interest, since they affect the accuracy of the surface finish, the chip thickness removal, and cutting forces exciting the machine tool. In the following, the basic principles of modal analysis are reviewed using the 2-DOF system shown in Figure 3.13 as an example [54].

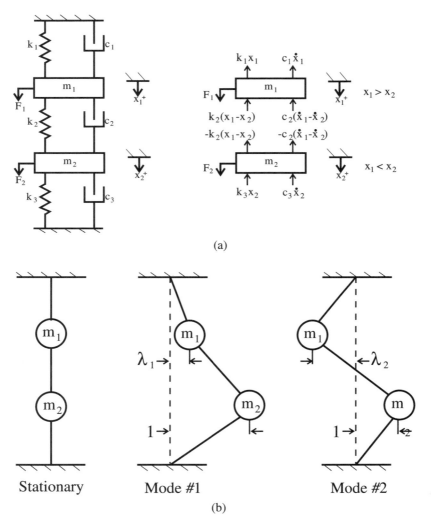

Figure 3.13: a) Mathematical model of a sample 2-DOF system; b) mode shapes of a 2-DOF system.

From Newton's second law, the equations of motion for masses m_1 and m_2 in local or measurement coordinates (x_1, x_2) can be written as

$$m_1 \ddot{x}_1 = F_1 - c_1 \dot{x}_1 - c_2(\dot{x}_1 - \dot{x}_2) - k_1 x_1 - k_2(x_1 - x_2),$$

$$m_2 \ddot{x}_2 = F_2 - c_2(\dot{x}_2 - \dot{x}_1) - k_2(x_2 - x_1) - k_3 x_2 - c_3 \dot{x}_2.$$

Rearranging the equations in matrix form yields

$$[M_x]\{\ddot{x}\} + [C_x]\{\dot{x}\} + [K_x]\{x\} = \{F\}, \tag{3.20}$$

where the mass, stiffness, and damping matrices are

$$[M_x] = \begin{bmatrix} m_1 & 0 \\ 0 & m_2 \end{bmatrix}, \quad [K_x] = \begin{bmatrix} k_1 + k_2 & -k_2 \\ -k_2 & k_2 + k_3 \end{bmatrix}, \quad [C_x] = \begin{bmatrix} c_1 + c_2 & -c_2 \\ -c_2 & c_2 + c_3 \end{bmatrix},$$

respectively. The displacement vector $\{x\}$ and force vector $\{F\}$ are defined as

$$\{x\} = \begin{Bmatrix} x_1(t) \\ x_2(t) \end{Bmatrix}, \quad \{F\} = \begin{Bmatrix} F_1(t) \\ F_2(t) \end{Bmatrix}.$$

The solution to the preceeding set of differential equations is first obtained for an undamped free vibration case (i.e., $c_1 = c_2 = c_3 = 0$ and $\{F\} = \{0\}$):

$$[M_x]\{\ddot{x}\} + [K_x]\{x\} = \{0\}. \tag{3.21}$$

The undamped system has a general solution of

$$\{x(t)\} = \{X\} \sin(\omega t + \psi),$$

where $\{X\}$ and ψ are constants and ω is the natural frequency of the system. Substituting the displacement vector $(x\{t\})$ and its second derivative as an acceleration vector $\{\ddot{x}\} = -\omega^2 \{X\} \sin(\omega t + \psi)$, Eq. (3.21) becomes

$$([K_x] - \omega^2 [M_x])\{X\} = \{0\}, \tag{3.22}$$

or

$$\begin{bmatrix} k_1 + k_2 - \omega^2 m_1 & -k_2 \\ -k_2 & k_2 + k_3 - \omega^2 m_2 \end{bmatrix} \begin{Bmatrix} X_1 \\ X_2 \end{Bmatrix} = \begin{Bmatrix} 0 \\ 0 \end{Bmatrix}.$$

The determinant of these simultaneous algebraic equations must be zero for a nontrivial solution. Letting $s = \omega^2$, we have

$$\begin{vmatrix} k_1 + k_2 - s m_1 & -k_2 \\ -k_2 & k_2 + k_3 - s m_2 \end{vmatrix} = 0,$$

or

$$s^n + a_1 s^{n-1} + \cdots + a_n = 0, \tag{3.23}$$

where n is the system's number of degrees of freedom. For the 2-DOF system example,

$$s^2 - \left(\frac{k_1 + k_2}{m_1} + \frac{k_2 + k_3}{m_2}\right)s + \frac{k_1 k_2 + k_2 k_3 + k_1 k_3}{m_1 m_2} = 0.$$

This polynomial has two real values of s: $s_1 = \omega_{n1}^2$ and $s_2 = \omega_{n2}^2$, where ω_{n1} and ω_{n2} are the natural frequencies of the system. Superposing the contribution of each solution or mode, we get

$$\begin{Bmatrix} x_1(t) \\ x_2(t) \end{Bmatrix}_1 = \begin{Bmatrix} X_1 \\ X_2 \end{Bmatrix}_1 \sin(\omega_{n1}t + \Psi_1) + \begin{Bmatrix} X_1 \\ X_2 \end{Bmatrix}_2 \sin(\omega_{n2}t + \Psi_2), \tag{3.24}$$

where $\{P\}_{1,2} = \{X_1 \; X_2\}_{1,2}^T$ are the *eigenvectors* or *mode shapes* associated with the fundamental (ω_{n1}) and second natural frequencies (ω_{n2}), respectively. X_{ik} is the displacement of node i and Ψ_k is the phase contributed by the natural mode k. The solution of Eq. (3.22) gives only the ratio of amplitudes at each node. It is customary to normalize them with respect to a reference coordinate (i.e., X_2). Substituting ω_{n1} and ω_{n2} in Eq. (3.22) and rearranging yields

$$\left(\frac{X_1}{X_2}\right)_1 = \lambda_1 = \frac{k_2}{k_1 + k_2 - \omega_{n1}^2 m_1} = \frac{k_2 + k_3 - \omega_{n1}^2 m_2}{k_2},$$

$$\left(\frac{X_1}{X_2}\right)_2 = \lambda_2 = \frac{k_2}{k_1 + k_2 - \omega_{n2}^2 m_1} = \frac{k_2 + k_3 - \omega_{n2}^2 m_2}{k_2}.$$

Substituting $\lambda_{1,2} = (X_1/X_2)_{1,2}$ and setting $X_{21} = Q_1$ and $X_{22} = Q_2$ into the transient vibrations (Eq. 3.24) gives

$$\begin{Bmatrix} x_1(t) \\ x_2(t) \end{Bmatrix} = \begin{bmatrix} \lambda_1 & \lambda_2 \\ 1 & 1 \end{bmatrix} \begin{Bmatrix} Q_1 \sin(\omega_{n1}t + \Psi_1) \\ Q_2 \sin(\omega_{n2}t + \Psi_2) \end{Bmatrix},$$

or in vector form

$$\{x(t)\} = [\{P\}_1 \; \{P\}_2] \begin{Bmatrix} q_1(t) \\ q_2(t) \end{Bmatrix} = [P]\{q(t)\}, \tag{3.25}$$

where $\{P\}_1 = \{\lambda_1 \; 1\}^T$ and q_1 are the first mode shape and modal displacement contributed by the first mode. The physical interpretation of a mode shape can be explained from Figure 3.13.b. The natural mode number causes one unit of displacement at mass m_2 and λ_1 units at mass m_1. $[P]$ is the complete modal matrix, which has dimension $[n \times n]$ for an n degree of freedom system. However, the modal matrix does not have to be square. The number of rows is equal to the number of coordinate points on the machine, and each column represents a mode.

Because the modes are orthogonal to each other, they have the following properties:

$$\{P\}_1^T [M_x]\{P\}_2 = 0,$$

whereas

$$\{P\}_1^T[M_x]\{P\}_1 = m_{q1},$$

where m_{q1} is the modal mass associated with the first mode. When the orthogonality principle is applied similarly to the remaining mode shapes, the local mass and stiffness matrices are transformed into modal coordinates:

$$\left.\begin{array}{l} [M_q] = [P]^T[M_x][P], \\ [K_q] = [P]^T[K_x][P]. \end{array}\right\} \tag{3.26}$$

The resulting modal mass ($[M_q]$) and modal stiffness ($[K_q]$) matrices are diagonal, and each diagonal element represents the modal mass or modal stiffness associated with a mode. Note that when the system has a proportional damping (i.e., $[C_x] = \alpha_1[M_x] + \alpha_2[K_x]$, where α_1 and α_2 are empirical constants), the transformed modal damping matrix $[C_q]$ is also diagonal, that is,

$$[C_q] = [P]^T[C_x][P]. \tag{3.27}$$

By applying modal transformations to Eq. (3.20), we can obtain the equations of motion in modal coordinates:

$$\begin{bmatrix} m_{q1} & 0 \\ 0 & m_{q2} \end{bmatrix} \begin{Bmatrix} \ddot{q}_1(t) \\ \ddot{q}_2(t) \end{Bmatrix} + \begin{bmatrix} c_{q1} & 0 \\ 0 & c_{q2} \end{bmatrix} \begin{Bmatrix} \dot{q}_1(t) \\ \dot{q}_2(t) \end{Bmatrix} + \begin{bmatrix} k_{q1} & 0 \\ 0 & k_{q2} \end{bmatrix} \begin{Bmatrix} q_1(t) \\ q_2(t) \end{Bmatrix} = \begin{Bmatrix} 0 \\ 0 \end{Bmatrix},$$

or, in vector notation,

$$[M_q]\{\ddot{q}\} + [C_q]\{\dot{q}\} + [K_q]\{q\} = \{0\}. \tag{3.28}$$

The equations of motion in modal coordinates are uncoupled and can be solved in a manner similar to that for the single degree of freedom systems. If we consider the free vibrations of the first mode as an example,

$$m_{q1}\ddot{q}_1 + c_{q1}\dot{q}_1 + k_{q1}q_1 = 0,$$

it has a solution of

$$q_1(t) = Q_1 e^{-\zeta \omega_{n1} t} \sin(\omega_{n1}\sqrt{1 - \zeta^2}t + \Psi_1),$$

where the modal damping ratio is $\zeta_1 = c_{q1}/(2\sqrt{k_{q1}m_{q1}})$ and Q_1 is found from initial conditions. After solving for the displacements in modal coordinates, the vibrations in local coordinates can be easily obtained by the modal transformation (3.25) (i.e., $\{x(t)\} = [P]\{q\}$). When the transformation is applied to the 2-DOF system example, the local displacements become

$$\begin{Bmatrix} x_1(t) \\ x_2(t) \end{Bmatrix} = \begin{bmatrix} \lambda_1 & \lambda_2 \\ 1 & 1 \end{bmatrix} \begin{Bmatrix} q_1 \\ q_2 \end{Bmatrix}.$$

By applying the same modal transformations to both sides of the equations of motion for forced vibrations (3.20), the force vector $\{F\}$ can be transformed to modal coordinates [53]:

$$\{R\} = [P]^{T}\{F\}. \tag{3.29}$$

The forced vibration expressions in modal coordinates become

$$[M_q]\{\ddot{q}\} + [C_q]\{\dot{q}\} + [K_q]\{q\} = \{R\}. \tag{3.30}$$

When the transformations are applied to the 2-DOF system example, we get

$$m_{q1}\ddot{q}_1 + c_{q1}\dot{q}_1 + k_{q1}q_1 = R_1,$$
$$m_{q2}\ddot{q}_2 + c_{q2}\dot{q}_2 + k_{q1}q_2 = R_2,$$

which are uncoupled differential equations and can be solved in a manner similar to that for single degree of freedom systems as explained earlier. The transfer functions in modal coordinates are expressed in a similar fashion to SDOF systems (Eq. 3.9) for each mode k:

$$|\Phi_{q,k}(\omega)| = \left|\frac{q_k}{R_k}\right| = \frac{1}{k_{qk}}\frac{1}{\sqrt{\left(1 - r_k^2\right)^2 + (2\zeta_k r_k)^2}},$$
$$\phi_k = \tan^{-1}\frac{-2\zeta_k r_k}{1 - r_k^2}, \tag{3.31}$$

where the frequency ratio is $r_k = \omega/\omega_{nk}$. For a 2-DOF example, the modal displacements can be expressed as

$$\begin{Bmatrix} q_1(t) \\ q_2(t) \end{Bmatrix} = \begin{bmatrix} \Phi_{q1} & 0 \\ 0 & \Phi_{q2} \end{bmatrix} \begin{Bmatrix} R_1 \\ R_2 \end{Bmatrix}.$$

Using a generalized matrix form, we have

$$\{q\} = [\Phi_q]\{R\}, \tag{3.32}$$

where Φ_q is the diagonal modal transfer function matrix. Substituting $\{R\} = [P]^T\{F\}$ and $\{x\} = [P]\{q\}$ into Eq. (3.32), we obtain the vibrations in local coordinates:

$$\{x\} = [P][\Phi_q][P]^T\{F\}, \tag{3.33}$$

or

$$\{x\} = \left(\sum_{k=1}^{n}\{P\}_k\{P\}_k^T\Phi_{q,k}\right)\{F\}, \tag{3.34}$$

where $\{P\}_k$ is the eigenvector for mode k, and n is the number of degrees of freedom. Thus the harmonically forced vibrations for each coordinate can be calculated using the mode shapes $\{P\}_k$, modal transfer functions Φ_{qk}, and the external force vector $\{F\}$. If the external force is $\{F\} = \{F_1 \sin(\omega_t)\ 0\}$, the measured vibrations at local coordinates x_1 and x_2 will be

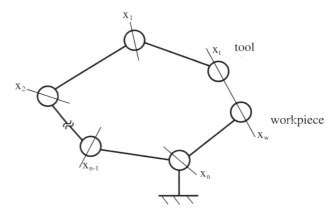

Figure 3.14: Transfer function between a tool and workpiece on the machine tools.

$$\frac{x_1(\omega)}{F_1(\omega)} = \lambda_1^2 \Phi_{q1} + \lambda_2^2 \Phi_{q2},$$

$$\frac{x_2(\omega)}{F_1(\omega)} = \lambda_1 \Phi_{q1} + \lambda_2 \Phi_{q2}.$$

3.4.5 Relative Transfer Function between Tool and Workpiece

Relative vibrations between the cutting tool and generated surface determine the accuracy of the machined workpiece and dynamic loading of the mechanical structures involved [45]. The cutting forces can be predicted for a given tool geometry, material constants, and cutting conditions as presented in the previous chapter. The cutting forces are applied on both the cutting tool and the workpiece with equal amplitudes but in opposite directions, and the resulting relative displacements between the two structures can be determined [53]. Consider the general machine tool structure shown in Figure 3.14. The relative displacement between the tool and workpiece is $(x_t - x_w)$, where x_t and x_w are displacements of the tool and workpiece, respectively. The cutting forces acting on the tool (F_t) and the workpiece (F_w) have the same magnitude of F_0 but have opposing directions (i.e., $F_t = -F_w$). The displacement and force vectors for an n-DOF system can be expressed in local coordinates as [53]

$$\{x\} = \{x_1, x_2, \dots, x_t;\quad x_w, \dots, x_n\},$$
$$\{F\} = \{0, 0, \dots, 1;\quad -1, \dots, 0\}F_0. \tag{3.35}$$

The equation of motion for the system is

$$[M_x]\{\ddot{x}\} + [C_x]\{\dot{x}\} + [K_x]\{x\} = \{F\},$$

where the local mass $[M_x]$, damping $[C_x]$, and stiffness $[K_x]$ matrices are square with dimensions $[n \times n]$, and the force vector $\{F\}$ has a dimension of $[n \times 1]$. The solution of the eigenvalue problem leads to a modal matrix $[P]$ with a dimension of $[n \times n]$:

$$[P] = \begin{bmatrix} P_{11} & P_{12} \dots P_{1t} & P_{1w} \dots P_{1n} \\ P_{21} & P_{22} \dots P_{2t} & P_{2w} \dots P_{2n} \\ \vdots & \vdots \quad \vdots & \vdots \quad \vdots \\ P_{t1} & P_{t2} \dots P_{tt} & P_{tw} \dots P_{tn} \\ P_{w1} & P_{w2} \dots P_{wt} & P_{ww} \dots P_{wn} \\ \vdots & \vdots \quad \vdots & \vdots \quad \vdots \\ P_{n1} & P_{n2} \dots P_{nt} & P_{nw} \dots P_{nn} \end{bmatrix},$$

or

$$[P] = [\{P\}_1, \{P\}_2, \dots, \{P\}_t, \{P\}_w, \dots \{P\}_n],$$

where each column represents a mode shape $\{P\}$ of the n-DOF system structure. From the modal coordinate transformation equation ($x = [P]\{q\}$), the tool and workpiece displacements can be calculated using the associated rows t and w of the above modal matrix:

$$\{x_t\} = \{P_{t1}\ P_{t2}\ \dots\ P_{tt}\ P_{tw} \dots P_{tn}\}\{q_1\ q_2 \dots q_t\ q_w \dots q_n\}^T,$$
$$\{x_w\} = \{P_{w1}\ P_{w2}\ \dots\ P_{wt}\ P_{ww} \dots P_{wn}\}\{q_1\ q_2 \dots q_t\ q_w \dots q_n\}^T. \tag{3.36}$$

Substituting the modal force ($\{R\} = [P]^T\{F\}$) into the modal displacement vector ($\{q\} = [\Phi_q]\{R\}$), and remembering that all elements of the force vector $\{F\}$ are zero except the two corresponding to coordinates x_t and x_w, yields

$$\{q\} = [\Phi_q]\begin{Bmatrix} P_{t1} - P_{w1} \\ P_{t2} - P_{w2} \\ \vdots \\ P_{tn} - P_{wn} \end{Bmatrix} F_0 = \begin{Bmatrix} \Phi_{q1}(P_{t1} - P_{w1}) \\ \Phi_{q2}(P_{t2} - P_{w2}) \\ \vdots \\ \Phi_{qn}(P_{tn} - P_{wn}) \end{Bmatrix} F_0.$$

Substitution of these modal displacements into Eq. (3.36) gives the local displacements of the tool and workpiece, respectively:

$$x_t = F_0 \sum_{i=1}^{n} \Phi_{qi} P_{ti}(P_{ti} - P_{wi}),$$
$$x_w = F_0 \sum_{i=1}^{n} \Phi_{qi} P_{wi}(P_{ti} - P_{wi}). \tag{3.37}$$

When the machine tool structure is excited by a harmonically varying force at the cutting point, the relative transfer function between the tool and workpiece becomes

$$\frac{x_t(\omega) - x_w(\omega)}{F_0(\omega)} = \sum_{i=1}^{n} \Phi_{qi}(P_{ti} - P_{wi})^2. \tag{3.38}$$

If the 2-DOF system is assumed to represent the tool and workpiece structures (i.e., $x_1 \equiv x_t$ and $x_2 \equiv x_w$), the force vector becomes $\{F\} = \{1 - 1\}F_0$. The relative transfer function between the tool and the workpiece in this particular example is found to be

$$\frac{x_1 - x_2}{F_0} = \Phi_{q1}(\lambda_1 - 1)^2 + \Phi_{q2}(\lambda_2 - 1)^2,$$

according to the outlined procedure.

3.5 EXPERIMENTAL MODAL ANALYSIS FOR MULTIDEGREE-OF-FREEDOM SYSTEMS

Transfer functions of existing multidegree-of-freedom (MDOF) systems are identified by structural dynamic tests. The physical machine is modeled by discrete lumped masses connected with linear and/or torsional springs. Good engineering judgment and experience must be applied to the observation of the physical structure and evaluation of its technical drawing.

The most effective excitation instruments are electromagnetic or electrohydraulic shakers. These are capable of delivering force over a wide frequency band containing the dominant natural modes of the tested structure. Alternatively, an impact hammer instrumented with a piezoelectric force transducer can be used (see Fig. 3.15). However, the impact hammer is

Figure 3.15: Measurement of transfer functions using a hammer instrumented with a force transducer and an accelerometer attached to a machine tool structure.

less effective in exciting the structure than are shakers, which can deliver force at random frequencies and amplitudes. The impact force delivered by the hammer has a short duration and can be treated as a narrow pulse or even as an impulse that has a flat Fourier spectrum. This means that a wide frequency range, which contains the hunted natural modes, is excited with the impulse force. The mass of the hammer and the impact tip material attached to the impact force sensor must be selected according to the mass, rigidity, and material of the structure being excited.

The resulting vibrations are measured with displacement, velocity, or acceleration sensors. Capacitance- or inductance-type displacement pickups are desirable to obtain static or low-frequency vibration signals in order to measure static flexibility. Because they are noncontact sensors, it can be difficult to mount them on different locations of the machine being tested. Recently developed laser displacement transducers are more practical to mount. The use of accelerometers is more common in measuring vibrations. The mass and frequency range of the accelerometers must be properly selected. Because they are mounted on the structure, their added mass may alter the measured natural frequencies. The bonding and interface materials between the structure and accelerometer must also be carefully selected to obtain accurate measurements.

The basics of experimental modal analysis are briefly introduced in the following, since all manufacturing engineers are exposed to dynamic problems in research or on manufacturing floors. Further reading and measurement experience are necessary to acquire an advanced knowledge in the area. The reference books from Ewins [55] and training courses provided by the manufacturers of Fourier analyzers and commercial modal analysis software systems are quite useful.

Concept of Residues. The transfer function of a SDOF system (3.7) can be expressed in the Laplace domain:

$$h(s) = \frac{X(s)}{F(s)} = \frac{1/m}{s^2 + 2\zeta\omega_n s + \omega_n^2}, \tag{3.39}$$

where $s^2 + 2\zeta\omega_n s + \omega_n^2$ is the characteristic equation of the system that has two complex conjugate roots,

$$s_1 = -\zeta\omega_n + j\omega_d, \qquad s_1^* = -\zeta\omega_n - j\omega_d.$$

The transfer function (3.39) can be expressed by its partial fraction expansion,

$$h(s) = \frac{r}{s - s_1} + \frac{r^*}{s - s_1^*} = \frac{\alpha + \beta s}{s^2 + 2\zeta\omega_n s + \omega_n^2}, \tag{3.40}$$

where the *residue* is

$$r = \sigma + j\nu, \qquad r^* = \sigma - j\nu \tag{3.41}$$

and the corresponding parameters are

$$\alpha = 2(\zeta\omega_n\sigma - \omega_d\nu), \qquad \beta = 2\sigma. \tag{3.42}$$

The residues may have real and imaginary parts depending on the damping and the number of modes in the system. However, the residues have distinct values for a SDOF system:

$$r = \lim_{s=s_1}(s - s_1)\frac{1/m}{(s - s_1)(s - s_1^*)} = \frac{1/m}{s_1 - s_1^*} = \frac{1/m}{2j\omega_d},$$

$$r^* = \lim_{s=s_1^*}(s - s_1^*)\frac{1/m}{(s - s_1)(s - s_1^*)} = \frac{1/m}{s_1^* - s_1} = -\frac{1/m}{2j\omega_d}.$$

Hence, the real part of the residues must be zero ($\sigma = 0$ and $\beta = 0$) for SDOF systems. Note that for a unit mass the residue must have a value of $r = 1/(2j\omega_d)$.

(Note that the transfer functions are alternatively represented with the following residue notation in the literature as well:

$$h(s) = \frac{r'}{2j(s - s_1)} + \frac{r'^*}{2j(s - s_1^*)} = \frac{\alpha' + \beta's}{s^2 + 2\zeta\omega_n s + \omega_n^2},$$ (3.43)

where the *residue* is

$$r' = \sigma' + j\nu', \qquad r'^* = \sigma' - j\nu'$$ (3.44)

and the corresponding parameters are

$$\alpha' = \zeta\omega_n\nu' - \omega_d\sigma', \qquad \beta = \nu'.$$

Hence, the imaginary part of the residues must be zero ($\nu' = 0$ and $\beta' = 0$) for SDOF systems. Note that for a unit mass the residue must have a value of $r' = 1/\omega_d$.

One can switch from one notation to the next with the following residue transformation:

$$r = r'/(2j) \rightarrow \sigma = \nu'/2, \qquad \nu = \sigma'/2.$$

Some of the commercial modal analysis packages use this notation.

Transfer Function of a MDOF System. The equation of motion for a MDOF system is represented in s domain by taking the Laplace transform of (3.20):

$$([M]s^2 + [C]s + [K])\{X(s)\} = \{F(s)\},$$ (3.45)

or

$$[B(s)]\{X(s)\} = \{F(s)\}.$$

The transfer function matrix of the MDOF system is then

$$[H(s)] = \frac{\{X(s)\}}{\{F(s)\}} = \frac{\mathrm{adj}[B(s)]}{\|[B(s)]\|},$$ (3.46)

where $\|[B(s)]\|$ is the characteristic equation, and the solution of $\|[B(s)]\| = 0$ gives the eigenvalues of the MDOF system. Note that the transfer function matrix $[H(s)]$ has $[n \times n]$ dimension for an n-DOF system, and all its elements have a common denominator of $\|[B(s)]\|$.

The transfer function of the example 2-DOF system is

$$[H(s)] = \begin{bmatrix} h_{11}(s) & h_{12}(s) \\ h_{21}(s) & h_{22}(s) \end{bmatrix},$$

where

$$h_{11}(s) = \left[\frac{r_{11,1}}{s - s_1} + \frac{r_{11,1}^*}{s - s_1^*}\right]_{\text{mode 1}} + \left[\frac{r_{11,2}}{s - s_2} + \frac{r_{11,2}^*}{s - s_2^*}\right]_{\text{mode 2}},$$

or

$$h_{11}(s) = \left[\frac{\alpha_{11,1} + \beta_{11,1}s}{s^2 + 2\zeta_1\omega_{n,1}s + \omega_{n,1}^2}\right]_{\text{mode 1}} + \left[\frac{\alpha_{11,2} + \beta_{11,2}s}{s^2 + 2\zeta_2\omega_{n,2}s + \omega_{n,2}^2}\right]_{\text{mode 2}}.$$ (3.47)

The elements h_{il} in the transfer function matrix are obtained from experimental measurements. When the 2-DOF structure is excited by force $\{F\} = \{F_1 \ F_2\}$ and the vibration is measured at point 1, we have

$$h_{11} = \frac{x_1}{F_1} \to F_2 = 0; \qquad h_{12} = \frac{x_1}{F_2} \to F_1 = 0.$$

For example, the accelerometer is attached to point 1, and the structure is hit with an instrumented hammer at point 1 only to measure h_{11} and is hit at point 2 to measure h_{12}. By attaching the accelerometer at point 2 and exciting the structure at points 1 and 2, the respective transfer function elements h_{21} and h_{22} are measured with Fourier analyzers. Note that all the elements have common denominators but different residues or numerators. The transfer function matrix is symmetric for linear systems (i.e., $h_{12} \equiv h_{21}$). It will be proven later that the measurement of only one row or column of the transfer function is sufficient to model the complete transfer function matrix using the symmetry and modal matrix properties. The measured transfer functions are stored in the frequency domain by the analyzers, usually in the form of real and imaginary components at each frequency, although the analyzers have built-in transformation routines to display the measurement data in time or other frequency domain representations. The measured frequency domain transfer function data (h_{il}) is transferred to a digital computer loaded with modal analysis software. For a given number of natural modes, the modal analysis system scans the transfer function data for the dominant maximum resonance amplitudes and the corresponding frequencies where the real part of the transfer function is zero. These frequencies are the natural frequencies of the system. The system then fits a curve to the data with a denominator having a $(2 \times n)$-order polynomial. With further numerical processing, the transfer function curve is fitted to n independent second-order differential equations as shown in (3.47). Thus, the numerical values of natural frequency, damping, and residue for each mode are estimated from the curve fitting [56]. In general form, the element in row i and column l of the transfer function matrix $[H(s)]$ is measured as

$$h_{il} = \sum_{k=1}^{n} \frac{\alpha_{il,k} + \beta_{il,k}s}{s^2 + 2\zeta_k \omega_{n,k} s + \omega_{n,k}^2}, \tag{3.48}$$

where $\omega_{d,k}$, $\omega_{n,k}$, and ζ_k are the damped and undamped natural frequencies and the modal damping ratio for mode k of the system, respectively. The frequency response of the structure can be obtained by replacing $s = j\omega$, where the excitation frequency ω can be scanned in a range covering all natural frequencies. The same equation can also be used in a time domain simulation of forced and chatter vibrations by simply using a bilinear approximation, $s = [2(1 - z^{-1})]/[\delta t(1 + z^{-1})]$ where δt is the digital integration time interval and z^{-1} is the backward time-shift operator (i.e., $z^{-1}x(t) = x(t - \delta t)$). The complete transfer function can be represented in the following matrix form:

$$[H(s)] = \sum_{k=1}^{n} \frac{[R]_k}{s^2 + 2\zeta_k \omega_{n,k} s + \omega_{n,k}^2}, \tag{3.49}$$

where each element in the $[n \times n]$-dimensional matrix $[R]_k = [\alpha + \beta s]_k$ reflects the residues of mode k at row i and column l.

The mode shapes of the system are found from the estimated residues. Remembering from (3.34) that the displacement vector can be expressed by its mode shapes and modal transfer functions,

$$\{x\} = \left(\sum_{k=1}^{n} \{P\}_k \{P\}_k^T \Phi_{q,k} \right) \{F\}.$$

Thus

$$[H(s)] = \sum_{k=1}^{n} \{P\}_k \{P\}_k^T \Phi_{q,k}. \tag{3.50}$$

The eigenvectors $\{P\}_k$ can be scaled arbitrarily as stated before. Combining Eqs. (3.49) and (3.50), we have

$$[H(s)] = \sum_{k=1}^{n} \frac{\{P\}_k \{P\}_k^T}{m_{q,k}} \frac{1}{s^2 + 2\zeta_k \omega_{n,k} s + \omega_{n,k}^2} = \sum_{k=1}^{n} \frac{[R]_k}{s^2 + 2\zeta_k \omega_{n,k} s + \omega_{n,k}^2}.$$

Note that the modal mass for mode k using the unscaled modal matrix is

$$m_{q,k} = \{P\}_k^T [M_x] \{P\}_k.$$

Thus $(\{P\}_k^T \{P\}_k)/m_{q,k}$ represents the normalization of each eigenvector with the square root of the modal mass (i.e., $\{u\}_k = \{P\}_k/\sqrt{m_{q,k}}$). The identified residues therefore have the following relationship with the mode shapes:

$$(\{P\}_k \{P\}_k^T)/m_{q,k} \equiv \{u\}_k \{u\}_k^T = [R]_k, \tag{3.51}$$

where $\{u_k\}$ corresponds to the normalized mode shape giving a unity modal mass. In other words, the mass is unity when the following transformation is used:

$$\{u\}_k^T [M_x] \{u\}_k = 1.$$

This procedure is a mathematically convenient way to simplify the identification of mode shapes, modal stiffness, and modal damping constants of the structure.

The residue matrix for a particular mode k can be expressed in the following general form:

$$[R]_k = \begin{bmatrix} u_1 u_1 & u_1 u_2 & \cdots & u_1 u_l & \cdots & u_1 u_n \\ u_2 u_1 & u_2 u_2 & \cdots & u_2 u_l & \cdots & u_2 u_n \\ \vdots & \vdots & \vdots & \vdots & \vdots & \\ u_l u_1 & u_l u_2 & \cdots & u_l u_l & \cdots & u_l u_n \\ \vdots & \vdots & \vdots & \vdots & \vdots & \\ u_n u_1 & u_n u_2 & \cdots & u_n u_l & \cdots & u_n u_n \end{bmatrix}_k.$$

If we take column or row l of the residue matrix for mode k, we have

$$\left\{ \begin{array}{c} R_{1l} \\ R_{2l} \\ \vdots \\ R_{ll} \\ \vdots \\ R_{nl} \end{array} \right\}_k = \left\{ \begin{array}{c} u_1 u_l \\ u_2 u_l \\ \vdots \\ u_l u_l \\ \vdots \\ u_n u_l \end{array} \right\}_k,$$

where $k = 1, 2, \ldots, n$ and for n number of modes. Starting first with the solution of u_l where the excitation and measurement points match, the mode shapes can be calculated from only one row or column of transfer function measurements. The mode shape vector for mode k is found as follows:

$$\left. \begin{array}{rcl} u_{l,k} & = & \sqrt{R_{ll,k}} \\[2mm] u_{1,k} & = & \dfrac{R_{1l,k}}{u_{l,k}} \\[3mm] u_{2,k} & = & \dfrac{R_{2l,k}}{u_{l,k}} \\[3mm] \vdots & = & \vdots \\[2mm] u_{n,k} & = & \dfrac{R_{nl,k}}{u_{l,k}} \end{array} \right\}. \tag{3.52}$$

This procedure is repeated for all modes to construct the full modal matrix of the structure. The modal matrix of the system consists of mode shapes in the columns of an $[n \times m]$ matrix,

$$[U] = [\{u\}_1 \ \{u\}_2 \ \ldots \ \{u\}_n], \tag{3.53}$$

where m is the total number of modes in the system and n is the number of measurement points or coordinates on the structure. Note that the modal matrix does not have to be a square matrix. For example, there may be two or three measurement points with a different number of modes. This depends on the structure and the number of measurement points used for vibration analysis. The identified transfer function can be used to analyze the general behavior of the machine tool under different loading conditions and to study machine tool stability for chatter vibrations during cutting as presented in the following section.

Note that since the elements of matrix $[R] = [\alpha + \beta s]$ are complex numbers, the resulting mode shapes will be complex and dependent on the modal frequency (ω_d). It is possible to obtain simplified real mode shapes:

$$R_{il,k} = \alpha_{il,k} + \beta_{il,k} s, \qquad s = j\omega_d.$$

For the mode shapes to be real, the imaginary part of $R_{il,k}$ must be zero (i.e., $\beta_{il,k} = 2\sigma_{il,k} = 0$) or the real part of the residues must be zero ($\sigma_{il,k} = 0$). This corresponds to structures having proportional damping, where the damping

$[C]$ is a linear combination of mass $[M]$ and stiffness $[K]$. The residues become $r = jv, r^* = -jv$, and $\alpha_{il,k} = 2\omega_{\mathrm{d},k}v_{il,k}$. For example, the modal parameter in Eq. (3.52) becomes $u_{l,k} = \sqrt{R_{ll}} = \sqrt{2\omega_{\mathrm{d}k}v_{lk}}$.

Example 1. A 2-DOF system shown in Figure 3.26 is analyzed experimentally. The masses lumped at the middle and the tip are $m_0 = 0.76$ kg each, the beam dimensions are $l = 450$ mm, $b = 25.4$ mm, and $h = 5$ mm, and the specific mass density of the steel is $\rho = 7,860$ kg/m^3. The mass of the beam becomes $m_{\mathrm{b}} = bhl\rho = 0.45$ kg. An accelerometer was used to measure the vibrations, and an impact hammer instrumented with a piezoelectric force transducer was used to hit the structure (see Fig. 3.15). Transducers are connected to amplifiers and power units to convert the measured charge signals to amplified voltage. The amplified force signal line is connected to the input channel, and the amplified accelerometer signal is connected to the output channel of a Fourier analyzer. The calibration factors of both sensors are entered into the analyzer to obtain correct measurement units. The transfer function h_{11} is measured by hitting the structure at point 1, where the accelerometer is mounted. The experiment is repeated ten times and an average transfer function measurement is accepted from the analyzer. The cross transfer function element h_{12} is measured by hitting structure at point 2 while the accelerometer is still mounted at point 1. The measured transfer functions (h_{11}, h_{12}) are transferred to a computer and processed by modal analysis software.

The modal analysis software provided the following information for the two modes:

Estimated	Mode 1	Mode 2
ω_{d} (rad/s)	53.0041	349.9659
ζ (%)	0.0113	0.0047
ω_{n} (Hz)	8.433	55.677
Residues for h_{11} (m/N)	1.8303e−3−j6.9526e−3	1.1546e−5−j1.4518e−4
Residues for h_{12} (m/N)	2.0721e−4−j3.0926e−3	4.6012e−5+j3.7015e−4

Note that when the measurements are made with an acceleromoter, residues must be scaled to displacement units before processing them (i.e., by dividing the residues by $\omega_{\mathrm{d}}^2 e^{j\pi}$ for each mode). By substituting $\alpha = 2(\zeta\omega_{\mathrm{n}}\sigma - \omega_{\mathrm{d}}v)$ and $\beta = 2\sigma$, the following direct (h_{11}) and cross (h_{12}) transfer functions are obtained:

$$h_{11}(s) = \frac{0.7392 + 0.00366s}{s^2 + 2\zeta_1\omega_{\mathrm{n}1}s + \omega_{\mathrm{n}1}^2} + \frac{0.1017 + 2.3092e - 05s}{s^2 + 2\zeta_2\omega_{\mathrm{n}2}s + \omega_{\mathrm{n}2}^2},$$

$$h_{12}(s) = \frac{0.3281 + 4.1442e - 04s}{s^2 + 2\zeta_1\omega_{\mathrm{n}1}s + \omega_{\mathrm{n}1}^2} + \frac{-0.2589 + 9.2024e - 05s}{s^2 + 2\zeta_2\omega_{\mathrm{n}2}s + \omega_{\mathrm{n}2}^2}.$$

By replacing $s = j\omega$ and sweeping all frequencies of interest, the curve-fitted transfer functions of the structure can be reconstructed. The complex mode shapes for the first mode can be extracted from the residues:

$$u_{11} = \sqrt{\alpha_{11,1} + \beta_{11,1}s}\big|_{s=j\omega_{d1}} = \sqrt{0.7392 + j0.00366 * 53.0041}$$
$$= 0.8670 + j0.1119,$$

$$u_{21} = \frac{\alpha_{12,1} + \beta_{12,1}s}{u_{11}}\bigg|_{s=j\omega_{d1}} = \frac{0.3281 + j4.1442e - 04 * 53.0041}{0.8670 + j0.1119}$$
$$= 0.3754 - j0.0231.$$

For the second mode, we obtain

$$u_{12} = \sqrt{\alpha_{11,2} + \beta_{11,2}s}\big|_{s=j\omega_{d2}} = \sqrt{-0.2589 + j2.3092e - 05 * 349.9659}$$
$$= 0.3191 + j0.0127,$$

$$u_{22} = \frac{\alpha_{12,2} + \beta_{12,2}s}{u_{12}}\bigg|_{s=j\omega_{d2}} = \frac{-0.2589 + j9.2024e - 05 * 349.9659}{0.3191 + j0.0127}$$
$$= -0.8062 + j0.1329.$$

The resulting complex modal matrix for unity modal mass is

$$[U] = \left[\left\{\begin{matrix} 0.8670 + j0.1119 \\ 0.3754 - j0.0231 \end{matrix}\right\}_1 \quad \left\{\begin{matrix} 0.3191 + j0.0127 \\ -0.8062 + j0.1329 \end{matrix}\right\}_2\right].$$

Note that the complex part of the numerator can be neglected (i.e., $\beta = 0$) for simplicity, which leads to real mode shapes and real mass, stiffness, and damping matrices.

Since both modal masses are unity (i.e., $m_{q,1} = m_{q,2} = 1$) the modal stiffness values are

$$k_{q,1} = \omega_{n,1}^2 = 53.0074^2 = 2,809\,\text{N/m},$$

$$k_{q,2} = \omega_{n,2}^2 = 349.4684^2 = 122,480\,\text{N/m},$$

and the modal damping constants are

$$c_{q,1} = 2\zeta_1\omega_{n,1} = 2 \times 0.0113 \times 53.0074 = 1.1928\,\text{N/ms}^{-1},$$

$$c_{q,2} = 2\zeta_2\omega_{n,2} = 2 \times 0.0047 \times 349.9697 = 3.2741\,\text{N/ms}^{-1}.$$

The transfer functions of two modes in modal coordinates are

$$\Phi_{q,1} = \frac{1}{m_{q,1}s^2 + c_{q,1}s + k_{q,1}} = \frac{1}{s^2 + 1.1928s + 2,809},$$

$$\Phi_{q,2} = \frac{1}{m_{q,2}s^2 + c_{q,2}s + k_{q,2}} = \frac{1}{s^2 + 3.2741s + 122,480},$$

and when substituted in Eq. (3.49), the transfer function matrix is found as

$$\left\{\begin{matrix} X_1(s) \\ X_2(s) \end{matrix}\right\} = \left[\begin{matrix} u_{11}^2\Phi_{q,1} + u_{12}^2\Phi_{q,2} & u_{11}u_{21}\Phi_{q,1} + u_{12}u_{22}\Phi_{q,2} \\ u_{11}u_{21}\Phi_{q,1} + u_{12}u_{22}\Phi_{q,2} & u_{21}^2\Phi_{q,1} + u_{22}^2\Phi_{q,2} \end{matrix}\right] \left\{\begin{matrix} F_1(s) \\ F_2(s) \end{matrix}\right\},$$

where $m_{q,1} = m_{q,2} = 1$, and real mode shapes (real(u)) are considered. The mass, damping, and stiffness matrices in local coordinates can be obtained from the inverse transformation using the modal matrix:

$$[M_x] = [U]^{-T}[I]U^{-1} = \begin{bmatrix} 1.1797 & -0.1018 \\ -0.1018 & 1.2732 \end{bmatrix} (\text{kg}).$$

The actual mass of the beam (m_b) can be distributed to points 1 and 2 as follows:

$$m_1 = m_0 + m_b/4 = 0.76 + 0.45/4 = 0.873 \, \text{kg},$$
$$m_2 = m_0 + m_b/2 = 0.76 + 0.45/2 = 0.985 \, \text{kg}.$$

The analytical actual masses are about 0.873 kg and 0.985 kg; the predicted masses in the diagonal locations of the mass matrix (1.1797 kg, 1.2732 kg) are slightly different, but the ratios are similar. The difference is due to errors in the measurement, noise, frequency resolution, accelerometer, and force sensor. The nonzero off-diagonal masses represent the presence of some dynamic coupling in the system:

$$[C_x] = [U]^{-T}[C_q]U^{-1} = \begin{bmatrix} 1.8447 & -1.1320 \\ -1.1320 & 3.8524 \end{bmatrix} (\text{N/ms}^{-1}),$$

$$[K_x] = [U]^{-T}[K_q]U^{-1} = \begin{bmatrix} 0.2847e+05 & -0.5839e+05 \\ -0.5839e+05 & 1.3776e+05 \end{bmatrix} (\text{N/m}).$$

The stiffness values are quite reasonable; they reflect the losses in the clamping of the beam on the ground and measurement errors with the accelerometer.

3.6   CHATTER VIBRATIONS IN CUTTING

Machine tool chatter vibrations result from a self-excitation mechanism in the generation of chip thickness during machining operations. One of the structural modes of the machine tool–workpiece system is excited by cutting forces initially. A wavy surface finish left during the previous revolution in turning, or by a previous tooth in milling, is removed during the succeeding revolution or tooth period, which also leaves a wavy surface owing to structural vibrations [57]. Depending on the phase shift between the two successive waves, the maximum chip thickness may exponentially grow while oscillating at a chatter frequency that is close to but not equal to a dominant structural mode in the system. The growing vibrations increase the cutting forces and may chip the tool and produce a poor, wavy surface finish. The self-excited chatter vibrations may be caused by mode coupling or regeneration of the chip thickness [58]. The mode coupling chatter occurs when there are vibrations in two directions in the plane of cut. The regenerative chatter results from phase differences between the vibration waves left on both sides of the chip and occurs earlier than the mode coupling chatter in most machining cases. Hence, the fundamentals of regenerative chatter vibrations are explained in the following section using a simple orthogonal cutting process as an example.

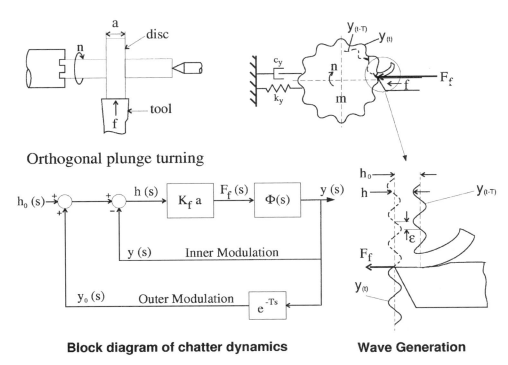

Figure 3.16: Regenerative chatter vibrations in orthogonal cutting.

3.6.1 Stability of Regenerative Chatter Vibrations in Orthogonal Cutting

Assume that a flat-faced orthogonal grooving tool is fed perpendicular to the axis of a cylindrical shaft held between the chuck and the tail stock center of a lathe (see Fig. 3.16). The shaft is flexible in the direction of feed, and the feed cutting force (F_f) causes it to vibrate. The initial surface of the shaft is smooth without waves during the first revolution, but the tool starts leaving a wavy surface behind because of the bending vibrations of the shaft in the feed direction y, which is in the direction of radial cutting force (F_f). When the second revolution starts, the surface has waves both inside the cut where the tool is cutting (i.e., inner modulation, $y(t)$) and on the outside surface of the cut owing to vibrations during the previous revolution of cut (i.e., outer modulation, $y(t - T)$). Hence, the resulting dynamic chip thickness is no longer constant but varies as a function of vibration frequency and the speed of the workpiece. The general dynamic chip thickness can be expressed as follows:

$$h(t) = h_0 - [y(t) - y(t - T)], \tag{3.54}$$

where h_0 is the intended chip thickness, which is equal to the feed rate of the machine, and $[y(t) - y(t - T)]$ is the dynamic chip thickness produced owing to vibrations at the present time t and one spindle revolution period (T) before. Assuming that the workpiece is approximated as a single degree of freedom system in the radial direction, the equation of motion of the system can be expressed as:

$$m_y\ddot{y}(t) + c_y\dot{y}(t) + k_y y(t) = F_f(t) = K_f a h(t)$$
$$= K_f a[h_0 + y(t - T) - y(t)], \qquad (3.55)$$

where the feed cutting force is proportional to the cutting constant in the feed direction (K_f), with width of cut a and the dynamic chip load $h(t)$. Since the forcing function on the right-hand side depends on the present and past solutions of vibrations $(y(t), y(t - T))$ on the left side of the equation, the chatter vibration expression is a delay differential equation. Furthermore, if the vibration is too large, [i.e., $y(t) - y(t - T) > h_0$], then the tool jumps out of cut, thus producing a zero chip thickness and zero cutting force. In addition, the influence of vibration marks left on the surface during the previous revolutions may further complicate the computation of exact chip thickness owing to tool jumping out of cut at various instances and revolutions (i.e., a multiple regenerative effect). The cutting constant K_f may also change depending on the magnitude of instantaneous chip thickness and the orientation of the vibrating tool or workpiece, which adds an additional difficulty in the dynamic cutting process. When the flank face of the tool rubs against the wavy surface left behind, additional process damping is added to the dynamic cutting process and this attenuates the chatter vibrations. Because the whole process is too complex and nonlinear to model correctly analytically, time domain, numerical methods are widely used to simulate the chatter vibrations in machining. However, a clear understanding of chatter stability is still important and is best explained using a linear stability theory. The stability of chatter vibrations is analyzed using linear theory by Tobias [59], Tlusty [57], and Merrit [60]. Nonlinearities such as the tool jumping out of cut [61], multiple regeneration, process damping, and nonlinear cutting constant are neglected in linear stability analysis.

The chatter vibration system can be represented by the block diagram shown in Figure 3.16, where the parameters of the dynamic cutting process are shown in the Laplace domain. Input to the system is the desired chip thickness h_0, and the output of the feedback system is the current vibration $y(t)$ left on the inner surface. In the Laplace domain, $y(s) = \mathcal{L}y(t)$, and the vibration imprinted on the outer surface during the previous revolution is $e^{-sT}y(s) = \mathcal{L}y(t - T)$, where T is the spindle period. The dynamic chip thickness in the Laplace domain is

$$h(s) = h_0 - y(s) + e^{-sT}y(s) = h_0 + (e^{-sT} - 1)y(s), \qquad (3.56)$$

which produces a dynamic cutting force of

$$F_f(s) = K_f a h(s). \qquad (3.57)$$

The cutting force excites the structure and produces the current vibrations

$$y(s) = F_f(s)\Phi(s) = K_f a h(s)\Phi(s), \qquad (3.58)$$

where the transfer function of the single degree of workpiece structure is

$$\Phi(s) = \frac{y(s)}{F_f(s)} = \frac{\omega_n^2}{k_y(s^2 + 2\zeta\omega_n s + \omega_n^2)}.$$

Substituting $y(s)$ into $h(s)$ yields

$$h(s) = h_0 + (e^{-sT} - 1)K_f a h(s)\Phi(s),$$

and the resulting transfer function between the dynamic and reference chip loads becomes

$$\frac{h(s)}{h_0(s)} = \frac{1}{1 + (1 - e^{-sT})K_f a \Phi(s)}. \tag{3.59}$$

The stability of this closed-loop transfer function is determined by the roots (s) of its characteristic equation, that is,

$$1 + (1 - e^{-sT})K_f a \Phi(s) = 0.$$

Let the root of the characteristic equation be $s = \sigma + j\omega_c$. If the real part of the root is positive ($\sigma > 0$), the time domain solution will have an exponential term with positive power (i.e., $e^{+|\sigma|t}$). The chatter vibrations will grow indefinitely, and the system will be unstable. A negative real root ($\sigma < 0$) will suppress the vibrations with time (i.e., $e^{-|\sigma|t}$), and the system is stable with chatter vibration–free cutting. When the real part is zero ($s = j\omega_c$), the system is critically stable, and the workpiece oscillates with constant vibration amplitude at chatter frequency ω_c. Note that the chatter vibration frequency does not equal the natural frequency of the structure, since the characteristic equation of the dynamic cutting process has additional terms beyond the structure's transfer function. However, the chatter vibration frequency is still close to the natural mode of the structure. For critical borderline stability analysis ($s = j\omega_c$), the characteristic function becomes

$$1 + (1 - e^{-j\omega_c T})K_f a_{\text{lim}}\Phi(j\omega_c) = 0, \tag{3.60}$$

where a_{lim} is the maximum axial depth of cut for chatter vibration–free machining. The transfer function can be partitioned into real and imaginary parts (i.e., $\Phi(j\omega_c) = G + jH$). Rearranging the characteristic equation with real and complex parts yields

$$\{1 + K_f a_{\text{lim}}[G(1 - \cos\omega_c T) - H\sin\omega_c T]\} \\ + j\{K_f a_{\text{lim}}[G\sin\omega_c T + H(1 - \cos\omega_c T)]\} = 0.$$

Both real and imaginary parts of the characteristic equation must be zero. If the imaginary part is considered first, then

$$G\sin\omega_c T + H(1 - \cos\omega_c T) = 0$$

and

$$\tan\psi = \frac{H(\omega_c)}{G(\omega_c)} = \frac{\sin\omega_c T}{\cos\omega_c T - 1}, \tag{3.61}$$

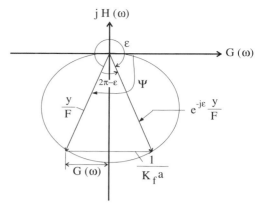

Figure 3.17: Calculation of phase angle ψ from the polar plots.

$\quad G > 0, H < 0 \rightarrow \psi = -\tan^{-1}|(H/G)|.$
$\quad G < 0, H < 0 \rightarrow \psi = -\pi + \tan^{-1}|(H/G)|.$
$\quad G < 0, H > 0 \rightarrow \psi = -\pi - \tan^{-1}|(H/G)|.$
$\quad G > 0, H > 0 \rightarrow \psi = -2\pi + \tan^{-1}|(H/G)|.$

where ψ is the phase shift of the structure's transfer function. Using the trigonometric identity $\cos \omega_c T = \cos^2(\omega_c T/2) - \sin^2(\omega_c T/2)$ and $\sin \omega_c T = 2 \sin(\omega_c T/2) \cos(\omega_c T/2)$, we have

$$\tan \psi = \frac{\cos(\omega_c T/2)}{-\sin(\omega_c T/2)} = \tan[(\omega_c T)/2 - (3\pi)/2]$$

and

$$\omega_c T = 3\pi + 2\psi \rightarrow \psi = \tan^{-1}\frac{H}{G}. \tag{3.62}$$

Note that calculating of the phase angle ψ from the transfer function must be correctly done with digital computers, as explained in Figure 3.17. The spindle speed (n [rev/s]) and the chatter vibration frequency (ω_c) have a relationship that affects the dynamic chip thickness. Let us assume that the chatter vibration frequency is ω_c[rad/s] or f_c[Hz]. The number of vibration waves left on the surface of the workpiece is

$$f_c \,[\text{Hz}] \cdot T \,[\text{s}] = \frac{f_c}{n} = k + \frac{\epsilon}{2\pi}, \tag{3.63}$$

where k is the integer number of waves and $\epsilon/2\pi$ is the fractional wave generated. The angle ϵ represents the phase difference between the inner and outer modulations. Note that if the spindle and vibration frequencies have an integer ratio, the phase difference between the inner and outer waves on the chip surface will be zero or 2π; hence the chip thickness will be constant despite the presence of vibrations. In this case, the inner ($y(t)$) and outer ($y(t - T)$) waves are parallel to each other. If the phase angle is not zero, the chip thickness changes continuously. Consider k integer number of full vibration cycles and the phase shift

$$2\pi f_c T = 2k\pi + \epsilon, \tag{3.64}$$

where the phase shift between the inner and outer waves is $\epsilon = 3\pi + 2\psi$. The corresponding spindle period (T[s]) and speed (n[rev/min]) is found to be

$$T = \frac{2k\pi + \epsilon}{2\pi f_c} \rightarrow n = \frac{60}{T}. \tag{3.65}$$

The critical axial depth of cut can be found by equating the real part of the characteristic equation to zero:

$$1 + K_f a_{lim}[G(1 - \cos \omega_c T - H \sin \omega_c T] = 0,$$

or

$$a_{lim} = \frac{-1}{K_f G[(1 - \cos \omega_c T - (H/G) \sin \omega_c T]}.$$

Substituting $H/G = (\sin \omega_c T)/(\cos \omega_c T - 1)$ and rearranging this equation yields

$$a_{lim} = \frac{-1}{2K_f G(\omega_c)}. \tag{3.66}$$

Note that since the depth of cut is a physical quantity, the solution is valid only for negative values of the real part of the transfer function ($G(\omega_c)$). The chatter vibrations may occur at any frequency where $G(\omega_c)$ is negative. If a_{lim} is selected using the minimum value of $G(\omega_c)$, the avoidance of chatter is guaranteed at any spindle speed. The expression indicates that the axial depth of cut is inversely proportional to the flexibility of the structure and to the cutting constant of the workpiece material. The harder the work material is, the larger the cutting constant K_f will be, thus reducing the axial depth of cut. Similarly, flexible machine tool or workpiece structures will also reduce the axial depth of cut or the productivity.

The above stability expression was first obtained by Tlusty [57]. Tobias [59] and Merrit [60] presented similar solutions. Tobias presented stability charts indicating chatter vibration–free spindle speeds and axial depth of cuts. Assuming that the transfer function of the structure at the cutting point (Φ) and cutting constant K_f are known or measured, the procedure for plotting the stability lobes can be summarized in the following:

- Select a chatter frequency (ω_c) at the negative real part of the transfer function.
- Calculate the phase angle of the structure at ω_c (Eq. 3.61).
- Calculate the critical depth of cut from Eq. (3.66).
- Calculate the spindle speed from Eq. (3.65) for each stability lobe $k = 0, 1, 2, \ldots$.
- Repeat the procedure by scanning the chatter frequencies around the natural frequency of the structure.

If the structure has multiple degrees of freedom, an oriented transfer function of the system in the direction of chip thickness must be considered for Φ. The computation of oriented transfer function was given in Section 3.4.2. In

Figure 3.18: Stability of a 2-DOF shaping process.

that case, the negative real part of the complete transfer function around all dominant modes must be scanned using the same procedure outlined for the orthogonal cutting process.

Example: Stability of a Shaping System with two Degrees of Freedom. A shaping process with a two degree of freedom system is shown in Figure 3.18. The system flexibilities in directions x_1 and x_2 are given as

$$x_1 \leftarrow \omega_{n1} = 250\,\text{Hz}, \quad \zeta_1 = 1.2\%, \quad k_1 = 2.26 \times 10^8\,\text{N/m},$$
$$x_2 \leftarrow \omega_{n2} = 150\,\text{Hz}, \quad \zeta_2 = 1.0\%, \quad k_2 = 2.13 \times 10^8\,\text{N/m}.$$

The cutting force is $F_y = K_f a h(t)$, where the cutting constant $K_f = 1,000\,\text{MPa}$. The flexibilities are oriented with $\theta_1 = 30°$ and $\theta_2 = -45°$ from the y axis. Plot the stability lobe of the system.

Answer. The real part of the oriented transfer function between the displacement in the y direction and the cutting force F_y is

$$\Re\Phi(j\omega) = G(j\omega) = \frac{y(j\omega)}{F_y(j\omega)}$$

$$= \cos^2\theta_1 \frac{1 - r_1^2}{k_1\left[\left(1 - r_1^2\right)^2 + (2\zeta_1 r_1)^2\right]} + \cos^2\theta_2 \frac{1 - r_2^2}{k_2\left[\left(1 - r_2^2\right)^2 + (2\zeta_2 r_2)^2\right]},$$

where $r_{1,2} = \omega/\omega_{n1,2}$. The oriented transfer function and the corresponding stability lobes are given in Figure 3.18.

3.7 ANALYTICAL PREDICTION OF CHATTER VIBRATIONS IN MILLING

The rotating cutting force and chip thickness directions and intermittent cutting periods complicate the application of orthogonal chatter theory to milling operations. The following analytical chatter prediction model was presented by Altintas and Budak [62, 63], who provide practical guidance to machine tool users for optimal process planning of depth of cuts and spindle speeds in milling operations.

3.7.1 Dynamic Milling Model

Milling cutters can be considered to have two orthogonal degrees of freedom as shown in Figure 3.19. The cutter is assumed to have N number of teeth with a zero helix angle. The cutting forces excite the structure in the feed (X) and normal (Y) directions, causing dynamic displacements x and y, respectively. The dynamic displacements are carried to rotating tooth number j in the radial or chip thickness direction with the coordinate transformation of $v_j = -x \sin\phi_j - y \cos\phi_j$, where ϕ_j is the instantaneous angular immersion of tooth j measured clockwise from the normal (Y) axis. If the spindle rotates at an angular speed of $\Omega(\text{rad/s})$, the immersion angle varies with time as $\phi_j(t) = \Omega t$. The resulting chip thickness consists of a static part $(s_t \sin\phi_j)$, attributed to rigid body motion of the cutter, and a dynamic component caused by the vibrations of the tool at the present and previous tooth periods. Since the chip thickness is measured in the radial direction (v_j), the total chip load can be expressed by

$$h(\phi_j) = [s_t \sin\phi_j + (v_{j,0} - v_j)]g(\phi_j), \tag{3.67}$$

where s_t is the feed rate per tooth and $(v_{j,0}, v_j)$ are the dynamic displacements of the cutter at the previous and present tooth periods, respectively. The function $g(\phi_j)$ is a unit step function that determines whether the tooth is in or out of

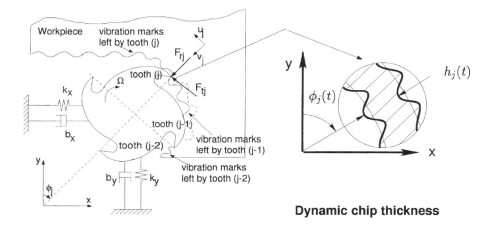

Dynamic chip thickness

End milling system

| **Static chip thickness** | **-** | **Vibration at present tooth period** | **+** | **Vibration at previous tooth period** |

$$h_j(t) = s_t \sin \phi_j(t) - \left[- x(t) \sin \phi_j(t) - y(t) \cos \phi_j(t) \right] + \left[-x(t-T) \sin \phi_j(t) - y(t-T) \cos \phi_j(t) \right]$$

Figure 3.19: Self-excited vibrations in 2-DOF milling system.

cut, that is,

$$\left. \begin{aligned} g(\phi_j) &= 1 \leftarrow \phi_{st} < \phi_j < \phi_{ex}, \\ g(\phi_j) &= 0 \leftarrow \phi_j < \phi_{st} \text{ or } \phi_j > \phi_{ex}, \end{aligned} \right\} \tag{3.68}$$

where ϕ_{st} and ϕ_{ex} are the start and exit immersion angles of the cutter to and from the cut, respectively. Henceforth, the static component of the chip thickness ($s_t \sin \phi_j$) is dropped from the expressions because it does not contribute to the dynamic chip load regeneration mechanism. Substituting v_j into (3.67) yields

$$h(\phi_j) = [\Delta x \sin \phi_j + \Delta y \cos \phi_j] g(\phi_j), \tag{3.69}$$

where $\Delta x = x - x_0$ and $\Delta y = y - y_0$. Here (x, y) and (x_0, y_0) represent the dynamic displacements of the cutter structure at the present and previous tooth periods, respectively. The tangential (F_{tj}) and radial (F_{rj}) cutting forces acting on the tooth j are proportional to the axial depth of cut (a) and chip thickness (h):

$$F_{tj} = K_t a h(\phi_j), \qquad F_{rj} = K_r F_{tj}, \tag{3.70}$$

where cutting coefficients K_t and K_r are constant. Resolving the cutting forces in the x and y directions,

$$\begin{aligned} F_{xj} &= -F_{tj} \cos\phi_j - F_{rj} \sin\phi_j, \\ F_{yj} &= +F_{tj} \sin\phi_j - F_{rj} \cos\phi_j, \end{aligned} \tag{3.71}$$

and summing the cutting forces contributed by all teeth, we can write the total dynamic milling forces acting on the cutter as

$$F_x = \sum_{j=0}^{N-1} F_{x_j}(\phi_j); \qquad F_y = \sum_{j=0}^{N-1} F_{y_j}(\phi_j), \tag{3.72}$$

where $\phi_j = \phi + j\phi_p$ and cutter pitch angle is $\phi_p = 2\pi/N$. Substituting the chip thickness (3.69) and tooth forces (3.70) into (3.71) and rearranging the resulting expressions in matrix form yields

$$\begin{Bmatrix} F_x \\ F_y \end{Bmatrix} = \frac{1}{2} a K_t \begin{bmatrix} a_{xx} & a_{xy} \\ a_{yx} & a_{yy} \end{bmatrix} \begin{Bmatrix} \Delta x \\ \Delta y \end{Bmatrix}, \tag{3.73}$$

where time-varying *directional dynamic milling force coefficients* are given by

$$a_{xx} = \sum_{j=0}^{N-1} -g_j[\sin 2\phi_j + K_r(1 - \cos 2\phi_j)],$$

$$a_{xy} = \sum_{j=0}^{N-1} -g_j[(1 + \cos 2\phi_j) + K_r \sin 2\phi_j],$$

$$a_{yx} = \sum_{j=0}^{N-1} g_j[(1 - \cos 2\phi_j) - K_r \sin 2\phi_j],$$

$$a_{yy} = \sum_{j=0}^{N-1} g_j[\sin 2\phi_j - K_r(1 + \cos 2\phi_j)].$$

Considering that the angular position of the parameters changes with time and angular velocity, we can express Eq. (3.73) in time domain in a matrix form as [64, 65]

$$\{F(t)\} = \frac{1}{2} a K_t [A(t)]\{\Delta(t)\}. \tag{3.74}$$

As the cutter rotates, the directional factors vary with time, which is the fundamental difference between milling and operations such as turning where the direction of the force is constant. However, like the milling forces, $[A(t)]$ is periodic at tooth passing frequency $\omega = N\Omega$ or tooth period $T = 2\pi/\omega$. Thus it can be expanded into a Fourier series:

$$[A(t)] = \sum_{r=-\infty}^{\infty} [A_r]e^{ir\omega t}, \tag{3.75}$$

$$[A_r] = \frac{1}{T}\int_0^T [A(t)]e^{-ir\omega t}\,dt.$$

The number of harmonics (r) of the tooth passing frequency (ω) to be considered for an accurate reconstruction of $[A(t)]$ depends on the immersion conditions and on the number of teeth in cut. If the most simplistic approximation, the average component of the Fourier series expansion, is considered (i.e., $r = 0$) then

$$[A_0] = \frac{1}{T}\int_0^T [A(t)]\,dt. \tag{3.76}$$

Since $[A_0]$ is valid only between the entry (ϕ_{st}) and exit (ϕ_{ex}) angles of the cutter (i.e., $g_j(\phi_j) = 1$), and $\phi_j = \Omega t$ and $\phi_p = \Omega T$, it becomes equal to the average value of $[A(t)]$ at cutter pitch angle $\phi_p = 2\pi/N$:

$$[A(0)] = \frac{1}{\phi_p}\int_{\phi_{st}}^{\phi_{ex}} [A(\phi)]\,d\phi = \frac{N}{2\pi}\begin{bmatrix} \alpha_{xx} & \alpha_{xy} \\ \alpha_{yx} & \alpha_{yy} \end{bmatrix}, \tag{3.77}$$

where the integrated functions are given as

$$\alpha_{xx} = \frac{1}{2}\Big[\cos 2\phi - 2K_r\phi + K_r \sin 2\phi\Big]_{\phi_{st}}^{\phi_{ex}},$$

$$\alpha_{xy} = \frac{1}{2}\Big[-\sin 2\phi - 2\phi + K_r \cos 2\phi\Big]_{\phi_{st}}^{\phi_{ex}},$$

$$\alpha_{yx} = \frac{1}{2}\Big[-\sin 2\phi + 2\phi + K_r \cos 2\phi\Big]_{\phi_{st}}^{\phi_{ex}},$$

$$\alpha_{yy} = \frac{1}{2}\Big[-\cos 2\phi - 2K_r\phi - K_r \sin 2\phi\Big]_{\phi_{st}}^{\phi_{ex}}.$$

The average directional factors are dependent on the radial cutting constant (K_r) and the width of cut bound by entry (ϕ_{st}) and exit (ϕ_{ex}) angles as illustrated in Figure 3.20. The dynamic milling expression (3.74) is reduced to the following:

$$\{F(t)\} = \frac{1}{2}aK_t[A_0]\{\Delta(t)\}, \tag{3.78}$$

where $[A_0]$ is the time-invariant but immersion-dependent directional cutting coefficient matrix. Since the average cutting force per tooth period is independent of the helix angle, $[A_0]$ is valid for helical end mills as well.

It may be argued that the use of the average Fourier coefficient $[A_0]$ may not be sufficient in the stability analysis. When the width of cut is small, the milling force waveforms are narrow and intermittent. Such waveforms have strong harmonic components in addition to the average value. When the author's group considered the high harmonics using iterative solutions, they noticed that

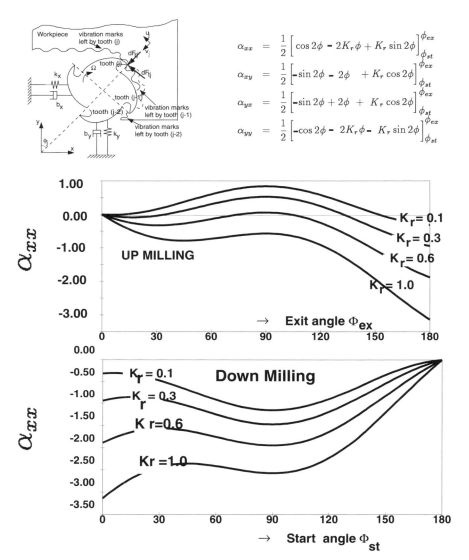

$$\alpha_{xx} = \frac{1}{2}\left[\cos 2\phi - 2K_r\phi + K_r\sin 2\phi\right]_{\phi_{st}}^{\phi_{ex}}$$

$$\alpha_{xy} = \frac{1}{2}\left[-\sin 2\phi - 2\phi + K_r\cos 2\phi\right]_{\phi_{st}}^{\phi_{ex}}$$

$$\alpha_{yx} = \frac{1}{2}\left[-\sin 2\phi + 2\phi + K_r\cos 2\phi\right]_{\phi_{st}}^{\phi_{ex}}$$

$$\alpha_{yy} = \frac{1}{2}\left[-\cos 2\phi - 2K_r\phi - K_r\sin 2\phi\right]_{\phi_{st}}^{\phi_{ex}}$$

Figure 3.20: The influence of radial cutting constant (K_r) and immersion (ϕ_{st}, ϕ_{ex}) on average directional factors.

the stability results did not improve even at very low immersion cuts. Physical analysis of the process illustrated in Figure 3.21 revealed that the high harmonics of the forcing function (i.e., milling forces) are low pass filtered within the dynamic cutting process. The process is explained from Figure 3.21 as follows. If chatter vibrations are present in a milling operation, the regenerative vibration spectra is dominated by the chatter frequency ω_c. The regenerative vibrations (i.e., *dynamic chip loads*) are multiplied by the immersion-dependent directional factors, which have an average value and additional strengths at the harmonics of tooth passing frequency (ω_T, $2\omega_T$, ...). The resulting dynamic milling forces have strengths at chatter frequency (ω_c) and at tooth passing harmonics

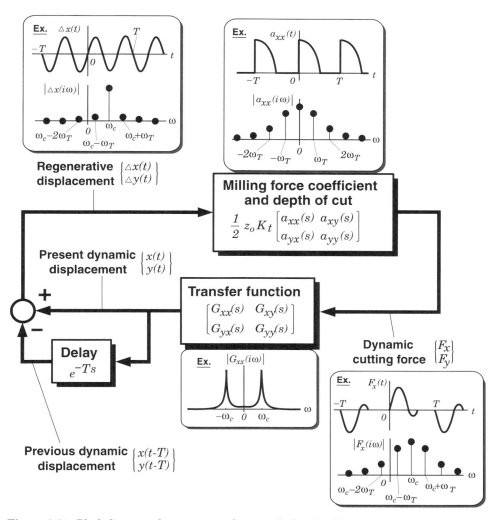

Figure 3.21: Block diagram of regenerative chatter in ball end milling process.

away from the chatter frequency ($\omega_c + i\omega_T$, $i = \pm 1, \pm 2, \ldots$). The dynamic cutting force excites the transfer function of the structure whose dominant mode is very close to the chatter frequency ω_c. The transfer function does not have any strength away from the natural mode that is causing the chatter; hence it low pass filters or attenuates the dynamic cutting force harmonics away from the chatter frequency. The chatter loop is closed when the structure leaves new chatter vibration marks on the present and previous surface cut during milling. However, if the structure has closely spaced modes, and if these modes happen to be distributed at tooth passing frequency intervals, the use of the average Fourier coefficient $[A_0]$ may not be sufficient for accurate prediction of chatter stability. In such cases, higher harmonics of the directional factors must be considered, and the resulting stability expression shown in the following section must be solved with numerical iterations.

3.7.2 Chatter Stability Lobes

The transfer function matrix ($[\Phi(i\omega)]$) identified at the cutter–workpiece contact zone is

$$[\Phi(i\omega)] = \begin{bmatrix} \Phi_{xx}(i\omega) & \Phi_{xy}(i\omega) \\ \Phi_{yx}(i\omega) & \Phi_{yy}(i\omega) \end{bmatrix}, \tag{3.79}$$

where $\Phi_{xx}(i\omega)$ and $\Phi_{yy}(i\omega)$ are the direct transfer functions in the x and y directions, and where $\Phi_{xy}(i\omega)$ and $\Phi_{yx}(i\omega)$ are the cross transfer functions. The vibration vectors at the present time (t) and previous tooth period ($t - T$) are defined as

$$\{r\} = \{x(t)\ y(t)\}^T\ ; \qquad \{r_0\} = \{x(t - T)\ y(t - T)\}^T.$$

Describing the vibrations at the chatter frequency ω_c in the frequency domain using harmonic functions, we write

$$\begin{aligned} \{r(i\omega_c)\} &= [\Phi(i\omega)]\{F\}e^{i\omega_c t}, \\ \{r_0(i\omega_c)\} &= e^{-i\omega_c T}\{r(i\omega_c)\}, \end{aligned} \tag{3.80}$$

and substituting $\{\Delta\} = \{(x - x_0)(y - y_0)\}^T$ gives

$$\begin{aligned} \{\Delta(i\omega_c)\} &= \{r(i\omega_c)\} - \{r_0(i\omega_c)\} \\ &= [1 - e^{-i\omega_c T}]e^{i\omega_c t}[\Phi(i\omega_c)]\{F\}, \end{aligned}$$

where $\omega_c T$ is the phase delay between the vibrations at successive tooth periods T. Substituting $\{\Delta(i\omega_c)\}$ into the dynamic milling equation (3.78) gives

$$\{F\}e^{i\omega_c t} = \frac{1}{2}aK_t[1 - e^{-i\omega_c T}][A_0][\Phi(i\omega_c)]\{F\}e^{i\omega_c t},$$

which has a nontrivial solution if its determinant is zero:

$$\det\left[[I] - \frac{1}{2}K_t a(1 - e^{-i\omega_c T})[A_0][\Phi(i\omega_c)] \right] = 0.$$

This is the characteristic equation of the closed loop dynamic milling system. The notation is further simplified by defining the oriented transfer function matrix as

$$[\Phi_0(i\omega_c)] = \begin{bmatrix} \alpha_{xx}\Phi_{xx}(i\omega_c) + \alpha_{xy}\Phi_{yx}(i\omega_c) & \alpha_{xx}\Phi_{xy}(i\omega_c) + \alpha_{xy}\Phi_{yy}(i\omega_c) \\ \alpha_{yx}\Phi_{xx}(i\omega_c) + \alpha_{yy}\Phi_{yx}(i\omega_c) & \alpha_{yx}\Phi_{xy}(i\omega_c) + \alpha_{yy}\Phi_{yy}(i\omega_c) \end{bmatrix} \tag{3.81}$$

and the eigenvalue of the characteristic equation as

$$\Lambda = -\frac{N}{4\pi}aK_t(1 - e^{-i\omega_c T}). \tag{3.82}$$

The resulting characteristic equation becomes

$$\det[[I] + \Lambda[\Phi_0(i\omega_c)]] = 0. \tag{3.83}$$

The eigenvalue of Eq. (3.83) can easily be solved for a given chatter frequency ω_c, static cutting factors (K_t, K_r) (which can be stored as material-dependent quantities for any milling cutter geometry), radial immersion (ϕ_{st}, ϕ_{ex}), and transfer function of the structure (3.81). If two orthogonal degrees of freedom in feed (X) and normal (Y) directions are considered (i.e., $\Phi_{xy} = \Phi_{yx} = 0.0$), the characteristic equation becomes just a quadratic function

$$a_0 \Lambda^2 + a_1 \Lambda + 1 = 0, \tag{3.84}$$

where

$$a_0 = \Phi_{xx}(i\omega_c)\Phi_{yy}(i\omega_c)(\alpha_{xx}\alpha_{yy} - \alpha_{xy}\alpha_{yx}),$$
$$a_1 = \alpha_{xx}\Phi_{xx}(i\omega_c) + \alpha_{yy}\Phi_{yy}(i\omega_c).$$

Then, the eigenvalue Λ is obtained as

$$\Lambda = -\frac{1}{2a_0}\left(a_1 \pm \sqrt{a_1^2 - 4a_0}\right). \tag{3.85}$$

As long as the plane of cut (x, y) is considered, the characteristic equation is still a simple quadratic function regardless of the number of modes considered in the machine tool structure. Since the transfer functions are complex, the eigenvalue has a real and an imaginary part, $\Lambda = \Lambda_R + i\Lambda_I$. Substituting the eigenvalue and $e^{-i\omega_c T} = \cos \omega_c T - i \sin \omega_c T$ in Eq. (3.82) gives the critical axial depth of cut at chatter frequency ω_c:

$$\begin{aligned} a_{\lim} = -\frac{2\pi}{NK_t}&\left[\frac{\Lambda_R(1 - \cos \omega_c T) + \Lambda_I \sin \omega_c T}{(1 - \cos \omega_c T)}\right. \\ &\left.+ i\frac{\Lambda_I(1 - \cos \omega_c T) - \Lambda_R \sin \omega_c T}{(1 - \cos \omega_c T)}\right]. \end{aligned} \tag{3.86}$$

Since $a_{\lim}$ is a real number, the imaginary part of Eq. (3.86) must vanish:

$$\Lambda_I(1 - \cos \omega_c T) - \Lambda_R \sin \omega_c T = 0. \tag{3.87}$$

By substituting

$$\kappa = \frac{\Lambda_I}{\Lambda_R} = \frac{\sin \omega_c T}{1 - \cos \omega_c T} \tag{3.88}$$

into the real part of Eq. (3.86) (the imaginary part vanishes), the final expression for chatter-free axial depth of cut is found as

$$a_{\lim} = -\frac{2\pi \Lambda_R}{NK_t}(1 + \kappa^2). \tag{3.89}$$

Therefore, given the chatter frequency (ω_c), the chatter limit in terms of the axial depth of cut can directly be determined from Eq. (3.89).

The corresponding spindle speeds are also found in a manner similar to the chatter in orthogonal cutting presented in the previous section. From Eq. (3.88),

$$\kappa = \tan \psi = \frac{\cos(\omega_c T/2)}{\sin(\omega_c T/2)} = \tan[\pi/2 - (\omega_c T/2)] \tag{3.90}$$

and the phase shift of the eigenvalue is $\psi = \tan^{-1}\kappa$, and $\epsilon = \pi - 2\psi$ is the phase shift between the inner and outer modulations (present and previous vibration marks). Thus if k is the integer number of full vibration waves (i.e., lobes) imprinted on the cut arc, then

$$\omega_c T = \epsilon + 2k\pi. \tag{3.91}$$

Again, care must be taken in calculating the phase shift (ψ) from the real (Λ_R) and imaginary (Λ_I) parts of the eigenvalue. The spindle speed n(rev/min) is simply calculated by finding the tooth passing period $T(s)$,

$$T = \frac{1}{\omega_c}(\epsilon + 2k\pi) \rightarrow n = \frac{60}{NT}. \tag{3.92}$$

In summary, the transfer functions of the machine tool system are identified, and the dynamic cutting coefficients are evaluated from the derived equation (3.77) for a specified cutter, workpiece material, and radial immersion of the cut. Then the stability lobes are calculated as follows [62]:

- Select a chatter frequency from transfer functions around a dominant mode.
- Solve the eigenvalue equation (3.84).
- Calculate the critical depth of cut from Eq. (3.89).
- Calculate the spindle speed from Eq. (3.92) for each stability lobe $k = 0, 1, 2, \ldots$.
- Repeat the procedure by scanning the chatter frequencies around all dominant modes of the structure evident on the transfer functions.

Example: Face-Milling Cutter. A bull-nosed face-milling cutter with two circular inserts is used in milling aluminum alloy Al-7075. The diameter of the cutter body is 31.75 mm, and the radius of circular inserts is 4.7625 mm. The identified cutting constants are $K_t = 1{,}319.41\,[\mathrm{N/mm^2}]$ and $K_r = 788.83\,[\mathrm{N/mm^2}]$. The transfer function of the cutter mounted on the spindle is measured, and the modal parameters are identified in the feed (x) and normal (y) directions in Table 3.1.

The stability lobes of the cutter are shown in Figure 3.22 for a half-immersion down-milling operation. The stability lobes are predicted using time domain, numerical simulations, which consider all nonlinearities in the dynamic milling

TABLE 3.1. Identified Modal Parameters of a Bull-Nose Cutter Mounted on the Spindle of a Vertical Machining Center. The Transfer Function Units Are in [m/N]

Mode x	ω_{nx} [Hz]	ζ_x	$\sigma_x + j\nu_x$
1	452.77	0.123718	9.202966E−05−j1.862195E−04
2	1448.53	0.01651	−4.181562E−05−j3.043618E−04

Mode y	ω_{ny} [Hz]	ζ_y	$\sigma_y + j\nu_y$
1	516.17	0.0243	−2.3929E−06−j1.721539E−04
2	1407.64	0.0324	4.055052E−05−j3.618808E−04

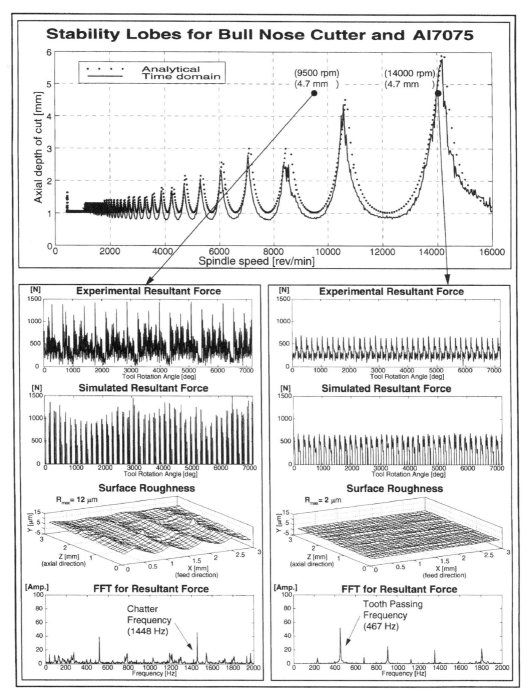

Figure 3.22: Stability of a bull-nosed face milling cutter with two circular inserts; the feed per tooth is $s_t = 0.05$ mm/tooth in sample measurenents and time domain simulations.

process such as tool jumping out, feed marks, and waves left by the previous cuts. The analytical and time domain solutions are in good agreement, although the analytical solution is linear and quite simple to implement in practice. Two cutting tests are conducted at $a = 4.7$ mm depth of cut. One is at the unstable spindle speed of $n = 9,500$ rev/min where severe chatter vibrations were observed, which are evident from force measurements and their Fourier spectrum. The chatter occurred close to the second mode of the structure at 1,448 Hz. When the spindle speed is increased to $n = 14,000$ rev/min, the chatter vibrations disappeared, and the spectrum of forces indicate that the dominance of forced vibration or tooth passing frequency was at 467 Hz. Time domain simulations demonstrate the presence of poor surface finish when chatter vibrations are present during milling.

Example: End Milling with a Flexible Cutter. A helical carbide end mill with 4 flutes, 19.05 mm diameter, and 10 degree rake is used in peripheral milling of aluminum alloy Al-7075. The identified cutting constants for the work material and cutter are found as $K_t = 796$ [N/mm^2] and $K_r = 0.212$. The transfer function of the cutter attached to the spindle is measured at the tool tip in both feed (x) and normal (y) directions with an instrumented hammer and accelerometer. The modal parameters of the structure converted to displacement domain are identified from modal analysis software and are given in Table 3.2.

TABLE 3.2. Identified Modal parameters of a Vertical Machining Center in x and y Directions. The Transfer Function Units Are in [m/N]

Mode x	ω_{nx} [Hz]	ζ_x	$\sigma_x + j\nu_x$
1	262.16	0.048	1.994606E−06−j 3.779091E−06
2	503.60	0.096	1.932445E−06−j 2.44616E−05
3	667.92	0.057	1.364547E−05−j 4.128093E−05
4	886.78	0.074	8.632813E−07−j 7.195716E−05
5	2201.5	0.015	−5.011467E−06−j 4.623555E−05
6	2799.4	0.027	−1.394105E−05−j 8.317728E−05
7	3837.7	0.071	−6.874543E−06−j 6.560925E−05
8	4923.6	0.019	−8.824935E−06−j 6.54481E−05
9	6038.7	0.025	2.564520E−06−j 1.986276E−05

Mode y	ω_{ny} [Hz]	ζ_y	$\sigma_y + j\nu_y$
1	285.53	0.021	9.797268E−07−j 5.959362E−07
2	587.8	0.089	1.349004E−05−j 4.544067E−06
3	749.6	0.027	−3.106831E−05−j 3.816475E−05
4	804.9	0.075	3.033849E−05−j 8.813244E−05
5	1573.7	0.027	1.283945E−06−j 8.678961E−06
6	2038.1	0.016	1.298625E−06−j 1.270846E−05
7	2303.3	0.220	9.518897E−07−j 3.750897E−05
8	2681.0	0.019	1.190834E−05−j 2.825781E−05
9	2870.5	0.014	−1.912932E−05−j 4.051860E−05
10	3838.8	0.060	−1.119589E−05−j 8.475725E−05
11	4928.9	0.017	−1.859880E−05−j 8.993226E−05
12	6073.4	0.016	−7.608392E−06−j 2.022991E−05

Figure 3.23: Experimentally measured and curve-fitted transfer functions of a vertical machining center at the tool tip.

The experimentally measured and reconstructed transfer functions are compared in Figure 3.23, which indicates the accuracy of identified modal parameters. Later the modal parameters are used to simulate the stability lobes for a half-immersion down-milling and slotting of Al-7075 alloy. These results are shown in Figure 3.24. The minimum stable depth of cut seems to be about $a_{lim} = 4$ mm in half-immersion up-milling and about $a_{lim} = 1.5$ mm in slotting. The ideal stability pockets correspond to a spindle speed $n = 11,800$ rev/min, where the maximum material can be removed without chatter. Note that there is hardly any stability pocket at speeds below $n = 5,000$ rev/min, which is due to the many tightly packed vibration waves at each tooth period. Since Al-7075 is not a hard material, it is clearly advisable to machine it at high speeds to enhance productivity. The stability lobes presented for this case have been experimentally proven on a machining center powered by a 15 kW spindle capable of 15,000 rev/min.

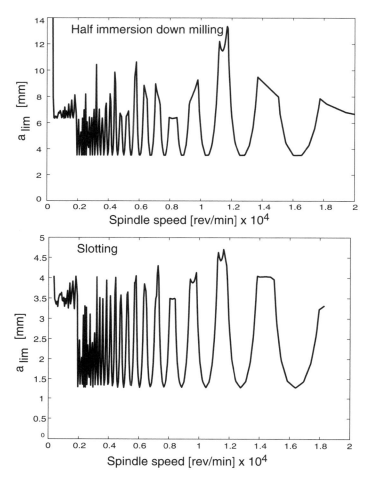

Figure 3.24: Analytically predicted stability lobes for helical end milling of aluminum alloy Al-7075.

3.8 PROBLEMS

1. A cylindrical AISI 4340 shaft with an elastic modulus of $E = 200$ GPa is placed between the chuck and the tail stock center of an engine lathe. The diameter of the shaft is $d = 30$ mm, the length of the bar is $L = 20$ mm, and the radial depth of cut is $a = 0.5$ mm. The radial cutting force (F_r) coefficient of the oblique turning tool is $K_r = 500$ MPa where $F_r = K_r ah$, and h is the chip load.

(a) Plot the dimensional errors left on the whole shaft as the tool is fed from the tail stock center toward the chuck. Assume that the tail stock and chuck supports are rigid.

(b) The stiffnesses of the chuck and tail stock are measured as $k_{sp} = 50{,}000$ N/mm and $k_{ts} = 30{,}000$ N/mm, respectively. Plot the dimensional errors again by including the stiffnesses of both supports.

Figure 3.25: Spring combinations.

2. A slender four-fluted high speed steel (HSS) end mill with a diameter of $d = 19.05\,\text{mm}$ and gauge length of $L = 100\,\text{mm}$ is used in half-immersion up- and down-milling of 7075 aluminum alloy. The tangential and radial force cutting constants are given as $K_t = 1200\,\text{MPa}$ and $K_r = 0.3$, where $F_t = K_t a h$ and $K_r = F_t/F_r$. The elastic modulus of HSS is $E = 204\,\text{GPa}$. Plot and compare the dimensional errors left on both up- and down-milled surfaces for the axial depth of cuts $a = 10, 20,$ and $30\,\text{mm}$.

3. Construct the stiffness matrices of the two structures shown in Figure 3.25 in local coordinates.

4. Two small rectangular steel parts, with identical masses of $m_0 = 0.76\,\text{kg}$, are attached to the middle and the end of a slender steel cantilever beam as shown in Figure 3.26. The dimensions of the steel bar are given as $l = 450\,\text{mm}$, $b = 25.4\,\text{mm}$, and $h = 5\,\text{mm}$. Young's modulus (E) and the mass density (ρ) of the steel are $E = 204\,\text{GPa}$ and $\rho = 7{,}860\,\text{kg/m}^3$. Assume that half of the bar mass is added to the middle and a quarter is added to the end point of the beam, and assume that the assembly is approximated as a 2-DOF system when it vibrates in its most flexible direction x. Develop a comprehensive computer program to solve the following:

(a) Obtain the stiffness matrix in local (x) coordinates using the flexibility method.

(b) Derive the equation of motion in local coordinates using Newton's law. Obtain the natural frequencies, modal matrix, and mode shapes. Plot the real mode shapes of the beam.

(c) Assume that the beam is modeled by two torsional springs at the base and at the middle. Using the torsional design stiffness of the beams

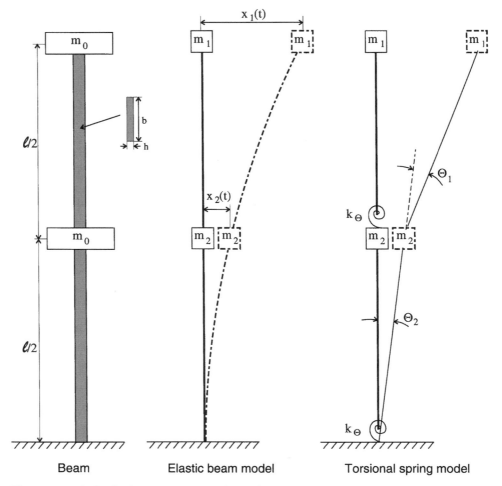

Figure 3.26: A slender beam carrying two lumped masses.

at these points, express the equation of motion using the Lagrangian formulation. Use torsional displacements (θ_1, θ_2) as design coordinates. Solve for the natural frequencies and mode shapes. Comment on the differences between the solutions in local and design coordinates.

(d) Express the transformation matrix between the local (x) and design (θ) coordinates. Obtain local mass and stiffness matrices using design stiffness, design mass, and coordinate transformation matrices.

(e) Express the modal mass and modal stiffness matrices. Predict the free vibrations of masses 1 and 2 when the tip of the bar is displaced 1 mm and released.

5. Assume that the same beam given in the previous question is analyzed by an experimental modal analysis. The measurements are analyzed by an

experimental modal analysis software, and the identified parameters of the structure are given in Example 1.

(a) Estimate the local mass, damping, and stiffness matrices.

(b) Estimate the torsional design spring constants.

(c) Plot the direct transfer function at points 1 and 2 and the cross transfer function.

(d) Compare the transfer functions obtained from analytical and experimental modal analysis methods.

6. A three-fluted end mill is used in machining Al-7075 alloy. The tangential and radial cutting constants are given as $K_t = 600$ MPa and $K_r = 0.07$, respectively. The end mill is represented by two orthogonal modes, which are evaluated as follows:

$$\omega_{nx} = 593.75 \text{ Hz}, \quad \zeta_x = 3.9\%, \quad K_x = 5.59 \times 10^6 \text{ [N/m]},$$
$$\omega_{ny} = 675 \text{ Hz}, \quad \zeta_y = 3.5\%, \quad K_y = 5.71 \times 10^6 \text{ [N/m]}.$$

Plot the stability lobes of the milling system for full and half-immersion cut.

7. The transfer functions of a 127 mm long end mill with 100 mm diameter attached to the spindle are measured at three points in both feed (x) and normal (y) directions (see Fig. 3.27). The accelerometer was attached to point 2, and the hammer instrumented with an impact force sensor is applied to all three points. The measured transfer functions are curve fitted with a modal analysis software, and the modal parameters are given in Table 3.3.

Figure 3.27: A four-fluted long helical end mill attached to a spindle; the cutter has diameter $d = 100$ mm and length $l = 127$ mm.

(a) Identify the real modal matrix in both (x, y) directions.

(b) Plot the mode shapes in the y direction.

(c) Evaluate the direct transfer function at point 1 in directions x and y (i.e., $\Phi_{xx} = ?, \Phi_{yy} = ?$).

(d) An aircraft Al-7075 wing spar is down-milled with a width of cut $b = 10$ mm using the same cutter and material. The cutting constants are given as $K_t = 752$ MPa and $K_r = 0.3$. Evaluate the stability lobes by assuming the dynamics only at the tool tip (point 1). The cross transfer functions are assumed to be negligible. Plot the stability lobes and provide a short recommendation for selecting the most optimal cutting conditions.

TABLE 3.3. Model Parameters of an End Mill Attached to a Spindle

Model Parameters	X		Y	
	Model 1	Model 2	Model 1	Model 2
ω_n(Hz)	479.4	588.7	493.3	637.9
ζ	0.51	0.029	0.068	0.027
σ_{12} (1/kg)	$1.304945.10^{-5}$	$-2.725243.10^{-6}$	$1.532536.10^{-5}$	$7.659394.10^{-6}$
jv_{12} (1/kg)	$-2.467218.10^{-5}$	$-6.02617.10^{-5}$	$-7.680401.10^{-5}$	$-1.260438.10^{-5}$
σ_{22} (1/kg)	$6.117676.10^{-6}$	$-2.225907.10^{-6}$	$9.606246.10^{-6}$	$4.553267.10^{-6}$
jv_{22} (1/kg)	$-1.408889.10^{-5}$	$-4.572173.10^{-5}$	$-5.5804.10^{-5}$	$-7.4116.10^{-6}$
σ_{32} (1/kg)	$1.415152.10^{-6}$	$-2.237442.10^{-6}$	$3.953402.10^{-6}$	$1.35349.10^{-6}$
jv_{32} (1/kg)	$-1.094268.10^{-5}$	$-3.452623.10^{-5}$	$-3.913159.10^{-5}$	$-6.116283.10^{-6}$

8. A vertical milling machine was modeled by Tlusty and Moriwaki [53] as shown in Figure 3.28. $x_1, \ldots, x_5$ are local coordinates. Design coordinates are the rotational angles $\alpha_1, \ldots, \alpha_5$ with corresponding design torsional springs $\gamma_1, \ldots, \gamma_5$. The design coordinates (α_i) are relative type, which exist as relative rotations around axes $A_1, \ldots, A_5$. From the measurements and analysis made, the local stiffness, mass, and damping matrices are identified and given below.

$[K_x]$: Local Stiffness Matrix [m/N]

$+3.0383e+08$	$-1.4309e+08$	$+4.7365e+07$	$-2.6006e+08$	$+5.9897e+07$
$-1.4309e+08$	$+1.2182e+08$	$-3.7234e+07$	$+5.8508e+07$	$-7.9269e+06$
$+4.7365e+07$	$-3.7234e+07$	$+1.7730e+07$	$-2.7861e+07$	$-1.8665e+03$
$-2.6006e+08$	$+5.8508e+07$	$-2.7861e+07$	$+4.2966e+08$	$-2.0022e+08$
$+5.9897e+07$	$-7.9269e+06$	$-1.8665e+03$	$-2.0022e+08$	$+1.6220e+08$

$[M_x]$: Local Mass Matrix [kg]

$+7.3779e+01$	$-1.1939e+01$	$+4.3726e-05$	$+5.5505e-04$	$-7.3919e-04$
$-1.1939e+01$	$+3.9957e+01$	$+3.4192e+00$	$-2.9887e-04$	$+1.7549e-04$
$+4.3726e-05$	$+3.4192e+00$	$+2.7603e+01$	$-3.6039e-05$	$-2.2129e-04$
$+5.5505e-04$	$-2.9887e-04$	$-3.6039e-05$	$+4.8745e+01$	$-2.7705e+01$
$-7.3919e-04$	$+1.7549e-04$	$-2.2129e-04$	$-2.7705e+01$	$+4.7240e+02$

$[C_x]$: Local Damping Matrix [N/m/s]

$+8.6796e+03$	$-4.5561e+03$	$+1.1930e+03$	$-3.4018e+03$	$-1.3886e+03$
$-4.5561e+03$	$+5.5794e+03$	$-1.3945e+03$	$-5.1046e+02$	$+4.4820e+02$
$+1.1930e+03$	$-1.3945e+03$	$+1.1798e+03$	$-8.6024e+01$	$-1.0541e+03$
$-3.4018e+03$	$-5.1046e+02$	$-8.6024e+01$	$+9.8393e+03$	$-5.6767e+03$
$-1.3886e+03$	$+4.4820e+02$	$-1.0541e+03$	$-5.6767e+03$	$+1.0013e+04$

Figure 3.28: Modal analysis model of a vertical milling machine presented by Tlusty [53]. The measurements are given in millimeters.

(a) Compute eigenvalues, eigenvectors, and damping ratios of individual modes for the system. Draw diagrams showing mode shapes imposed on the sketch of the model of the machine. All the significant flexibilities of the machine occur in the directions $x_1, \ldots, x_5$, while the flexibilities in the other two directions perpendicular to x are neglected.

(b) Now consider the relative motion between tool x_{10} and work piece x_{11}. Assume that the table and knee are rigidly connected with the column and participate with x_5 in the freedom α_5. The tool is a rigid extension of the rigid connection between coordinates x_1 and x_2. With this in mind, replace the coordinate system $x_1, \ldots, x_5$ by the coordinate system $x_{10}, x_2, x_3, x_4, x_{11}$. Transform the local k_x, m_x, and c_x to this new coordinate system. Compute the relative transfer function between the points x_{10} and x_{11}. Plot the real part and Nyquist (i.e., polar) plot of the relative tool–workpiece transfer function.

(c) As an extension to the identification problem, write the transformation matrix $[C]$ between the local and design coordinates, and transform the local stiffness $[K_x]$ to $[K_\alpha]$ to find the design stiffness of the five torsional springs.

(d) Plot stability lobes for the first ten lobes for a milling cutter with four teeth and cutting constants given as $K_t = 2{,}000$ MPa, and $K_r = 0.5$. Consider half-immersion up- and down-milling operations.

CHAPTER FOUR

TECHNOLOGY OF MANUFACTURING AUTOMATION

4.1 INTRODUCTION

Numerically controlled (NC) machine tools were developed to fulfill the contour machining requirements of complex aircraft parts and forming dies. The first NC machine tool was developed by Parsons Company and MIT in 1952 [66]. The first-generation numerically controlled units used digital electronic circuits and did not contain any actual central processing unit; therefore they were called NC or *hardwired* NC machine tools. In the 1970s, *computer numerically controlled* (CNC) machine tools were developed with minicomputers used as control units. With the advances in electronics and computer technology, current CNC systems employ several high-performance microprocessors and programmable logical controllers that work in a parallel and coordinated fashion. Current CNC systems allow simultaneous servo position and velocity control of all the axis monitoring of controller and machine tool performance, online part programming with graphical assistance, in-process cutting process monitoring, and in-process part gauging for completely unmanned machining operations. Manufacturers offer most of these features as options.

4.2 COMPUTER NUMERICALLY CONTROLLED UNIT

A typical CNC machine tool has three fundamental units: the mechanical machine tool unit, power units (motors and power amplifiers), and the CNC unit. Here, a brief introduction of a CNC system from the user's point of view is presented.

4.2.1 Organization of a CNC Unit

A CNC unit of a machine tool consists of one or more central processing units (CPUs), input/output devices, operator interface devices, and programmable logical controllers.

The CNC unit may contain several CPUs, or microprocessors, depending on the tasks required by the machine tool. A very basic three-axis CNC milling machine requires the fine coordinated feeding velocity and position control of all three axes and the spindle speed simultaneously. Current CNC systems tend to employ multiple CPUs depending on the number of computation tasks. In such a multiprocessor-based CNC system, additional CPU modules can be added to the CNC to extend its intelligence and functions. For example, in a

Figure 4.1: A multiaxis CNC machining center.

machining center, such as the one shown in Figure 4.1, contouring control, pallet control, coordination of all eight axes, cutting process monitoring, and graphically assisted part programming can take place simultaneously owing to employment of several computers on its CNC system.

Each CNC unit has a keyboard and monitor for the operator, as well as floppy drive units and high-speed serial communication ports for loading NC programs. It must be noted that input devices such as floppy disk drives are very fragile in a typical metal cutting shop floor because of the presence of dust, metal chip particles, and cutting fluid. CNC coordinate measuring machines (CMMs) are used in clean air and temperature-controlled metrology rooms, and therefore they do not have the same constraints in the insulation of CNC units used in metal cutting shops. In a fairly computerized factory floor, it is common to have a direct communication link between the master computer of the factory and individual CNC units. NC programs, production schedules, and recording of production times are continuously electronically communicated between the CNC units and the master computer. Such systems are called *distributed numerically controlled (DNC)* systems. Although this eliminates the necessity of having input devices on CNC units, it can still be justified to have them on the CNC units to ensure stand-alone use of the machine tool in case of communication failures. CNC systems, equipped with a monitor and a keyboard, allow operators to edit NC programs on site. However, in an organized production floor, this is not recommended. The capital and operating costs of CNC machine tools are rather high, and their use for other than continuous metal removal may not be justified. The process planners in the engineering office test NC programs with the aid of a graphically animated tool path, before they are released to the production floor. The methods of NC programming and tool path verification are explained in the following sections.

There are miscellaneous logic functions, which have to be carried out routinely by the CNC system. For example, activating the cutting lubricant pump, turning the spindle clockwise (CW) or counterclockwise (CCW), touching travel limits of the guides, and tool changing functions require only coordinated ON/OFF type logic signals, which are carried out by the programmable logical control (PLC) units of the CNC systems.

Each CNC unit is equipped with emergency stop, shut-down, feed rate, and spindle speed override buttons and a manual jog and machine homing switches for easy use of the machine tool by the operators. The number of operator interface units depends very much on the type and the functions of the machine tool, as well as on the richness of the CNC software library.

4.2.2 CNC Executive

NC programs are written in an internationally recognized standard language. Each commercial CNC unit is expected to accept standard NC codes. The NC codes are transferred to the CNC unit in ASCII format using input devices. The CNC executive is the main system software that decodes the NC codes block by block and sends appropriate commands to physical control, computation, and PLC units of the CNC system. For example, a 10-mm distance to be traveled at a 200-mm/s feed velocity command can be translated as follows: The real time clock is set to generate 10,000 pulses at a rate of 200,000 pulses/s (1 pulse $= 0.001$ mm position). The position pulses (i.e., the discrete velocity commands) are directed to the indicated machine tool axis position control units by converting to their analog voltage equivalent (i.e., typically within ± 10 V range). The analog voltage is amplified by the power unit and fed to the axis drive motors to deliver the desired motion. Miscellaneous functions, such as spindle ON and tool change commands, are translated as Boolean logic signals ($+5$ V or -5 V) for PLC units.

The CNC executive software is expected to carry out NC functions in a logical order. For example, it must turn the spindle before a controlled feeding or machining action takes place. It must load the tool on the spindle before positioning the tool for machining, although such commands may be given in the same NC block.

4.3 CNC MACHINE TOOL AXIS CONVENTIONS

The Electronic Industries Association, in its standard RS-367-A, lists 14 different axis designations or types of motion. While common machine tools may have up to 5 axes, gear shaping and cutter grinding machine tools may have up to 14 axes of controlled motions.

Figure 4.2: CNC machine tool axis conventions.

The machine tools are programmed in a Cartesian coordinate system. The Z axis is always aligned with the direction of spindle. The primary X motion is normally parallel to the longest dimension of the primary machine tool table. The Y axis is normally parallel to the shortest dimension of the primary machine tool table. Characters A, B, and C designate angular motions around the X, Y, and Z axes respectively. Axis conventions of various machine tool configurations are shown in Figure 4.2.

4.4 NC PART PROGRAM STRUCTURE

An NC part program represents the machining sequences, or *blocks*, used to produce a desired component shape. Each block starts with the letter N followed by the block sequence number. One block of a typical NC program would look like

N0040 G91 X25.00 Y10.00 Z - 12.55 F150 S1100 T06 M03 M07.

The NC program consists of *blocks*, and each block consists of several *words*. A word starts with a character followed by numbers that represent a specific command for the machine tool. The word starting with G represents the *prepara-tory* functions, and the M words are the *miscellaneous* functions in the NC program. The words starting with F and S represent the feed and spindle speeds, respectively. T represents the tool number. x, y, and z letters followed by the scalars represent the motion lengths in the designated axes. The particular NC block given above is interpreted as follows by the CNC executive software: Load tool number 6 on the spindle (T06), rotate the spindle clockwise at 1,100 rev/min (S1100, M03) and turn the cutting fluid on (M07) before the motion starts. Move the machine tool in 25 mm, 10 mm, and 12.55 mm increments (G91) in the x, y, and z directions with the resultant feeding velocity of 150 mm/min along the tool path. N040 represents the 40th block sequence of the entire part program.

The list of commonly used NC words are:

N...: Block sequence number.
G...: Preparatory function.
X...: Primary X motion dimension.
Y...: Primary Y motion dimension.
Z...: Primary Z motion dimension.
U...: Secondary motion parallel to X axis.
V...: Secondary motion parallel to Y axis.
W...: Secondary motion parallel to Z axis.
A...: Angular dimension about X axis.
B...: Angular dimension about Y axis.
C...: Angular dimension about Z axis.
I...: Interpolation parameter or thread lead parallel to X.
J...: Interpolation parameter or thread lead parallel to Y.
K...: Interpolation parameter or thread lead parallel to Z.
F...: Feed word.
M...: Miscellaneous function.
S...: Spindle speed word.
T...: Tool number word.
R...: Rapid traverse dimension in Z axis.

Digital codes in a word may have to be written in a specific given format for some CNC units. However, in general, most control systems allow free formatted word fields. All NC words and their functions are given in the following according to ISO 1056 international standards.

Preparatory (G) Functions

- *G00:* Point-to-point positioning. This provides point-to-point positioning of the feed drives at a rapid traverse rate of the machine tool along an uncontrolled random path. It must be noted that each drive traverses at the rapid feeding velocity independently; therefore the programmer must be careful to avoid collision of the tool with the clamps or the workpiece.

- *G01:* Linear interpolation. The tool path velocity is kept constant at the given feed along the indicated straight line. The feed drives' velocities are coordinated to keep the tool along the straight line.
- *G02, G03:* Circular interpolation clockwise (*G*02) or counterclockwise (*G*03). An arc generated by the coordinated motion of the two axes in which the tool moves clockwise or counterclockwise, when viewing the plane of motion in the negative direction of the axis perpendicular to the plane of motion.
- *G04:* A timed delay of a programmed motion.
- *G07:* A controlled velocity increase and decrease to programmed rate starting immediately.
- *G17–G19:* Used to identify the plane for such functions as circular interpolation, cutter compensation, and other functions as required.
- *G21–G23:* Precision cornering mode selection for linear and circular interpolation.
- *G33:* Constant lead thread cutting.
- *G34, G35:* Increasing lead (*G*34) or decreasing lead (*G*35) thread cutting.
- *G40:* Command that will discontinue any active cutter compensation or offset.
- *G41, G42:* Cutter compensation left (*G*41), cutter compensation right (*G*42). Cutter on left (or right) side of work surface looking from cutter in the direction of relative cutter motion with displacement normal to the cutter path to adjust for the difference between actual and programmed cutter radius or diameters.
- *G70:* Mode for programming in imperial (inch) units.
- *G71:* Mode for programming in metric (mm) units.
- *G80:* Cancel the active fixed cycle.
- *G81–G89:* A preset series of operations that direct machine axis movement and/or cause spindle operation to complete fixed machining cycles such as boring, drilling, tapping, or combinations thereof. Fixed cycles are also called *canned cycles* by machine tool manufacturers.
- *G90:* Coordinate inputs are given in *absolute coordinates* from a fixed Cartesian coordinate center.
- *G91:* Coordinate inputs are given *incrementally* from a previous tool location.
- *G94:* The feeds are given in inches (or millimeters)/minute.
- *G95:* The feed rate is given in inches (or millimeters)/revolution.

Miscellaneous Functions

- *M00:* Program stop. Terminates further program execution after the completion of other commands in the block.
- *M01:* Optional stop if it is enabled by the operator. The program continues after the execution of a *continue* command by the operator.

- *M02:* End of program indicating completion of machining cycle. Stops spindle, coolant, and feed after the completion of all commands in the last NC block.
- *M03, M04:* Start spindle clockwise (*M*03) or counterclockwise (*M*04).
- *M05:* Spindle off.
- *M06:* Tool change.
- *M07, M08:* Cutting fluid ON (*M*07), OFF (*M*08).
- *M19:* Oriented spindle stop at a predetermined angular position.
- *M30:* End of program. It stops feed, spindle, and cutting fluid and rewinds the NC program to the beginning.
- *M49:* Prevents operator from overriding spindle and feed speeds.

It has to be noted that any G or M codes used in the NC program are active until their opposites are used, which overrides the previous ones.

4.5 MAIN PREPARATORY FUNCTIONS

G00 – Point-to-Point Positioning

The table (or cutter) is positioned from one point to another without having to coordinate the velocities of any of the moving axes. This mode is usually used at rapid traverse rate for drilling and tapping and for rapid positioning of the cutter without cutting.

Continuous Path Contouring

The table's position is continuously controlled to keep the cutter on a desired trajectory profile (i.e., slanted line, arc, or spline segment) during machining. This involves continuous manipulation of velocities of each axis involved during contour machining. Real time digital interpolation methods, which are explained in detail in the next chapter, are employed for continuous path machining. The two fundamental interpolation commands, which are provided on every commercial CNC unit, are explained below.

G01 – Linear Interpolation Code. The velocities of two axes are controlled to keep the tool on a straight path in a plane of motion. Figure 4.3 shows a line segment to be machined by a milling cutter. The coordinates of the end mill center for the beginning point P_1(10 mm, 12 mm) and end point P_2(60 mm, 37 mm) are given in the XY plane. To keep the end mill following the straight line (P_1P_2) at a given vector feed velocity, the linear interpolation command G01 must be used. The corresponding NC programs in absolute (G90) and incremental (G91) programming modes, respectively, are

    N0010 G90 G01 X60.0 Y37.0 F300

or

    N0010 G91 G01 X50.0 Y25.0 F300.

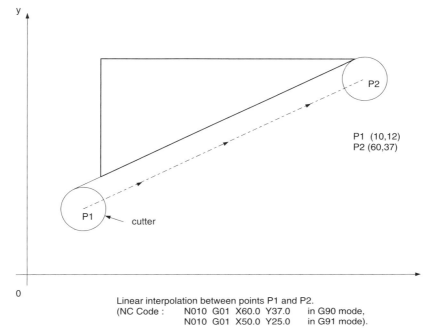

Figure 4.3: Linear interpolation.

G02, G03 – Circular Interpolation Codes. The velocities of two axes on a plane of motion are varied to keep the tool following the given arc at the specified feed velocity. There are two types of circular interpolation commands used in CNC systems. Some CNC systems require the coordinates of the arc center and arc's end point, while others need the radius of the arc and its end point. Figure 4.4 shows a sample arc segment to be contoured by an end mill. CNC assumes that the tool is located at the beginning point of the arc. Looking down the plane, or with respect to the previous tool motion, the arc contouring can be either clockwise (G02) or counterclockwise (G03). In Figure 4.4, the tool is required to move in a CCW direction (G03) at a constant contour feed velocity of f. If the CNC requires the radius (r_c) and the end point of the arc (P_2), the following NC program is used for the contour milling in absolute coordinates:

N010 G90 G03 $X x_2\ Y y_2\ R r_c\ F f$.

If a CNC requires the coordinates of the center point of the arc and the arc's end point, the NC program is

N010 G90 G03 $X x_2\ Y y_2\ I i_c\ J j_c\ F f$,

where the values of i_c and j_c are used to define the center of the arc with respect to the starting point. The following algorithm can be used to calculate the arc center offset:

$$i_c = x_c - x_1, \quad j_c = y_c - y_1.$$

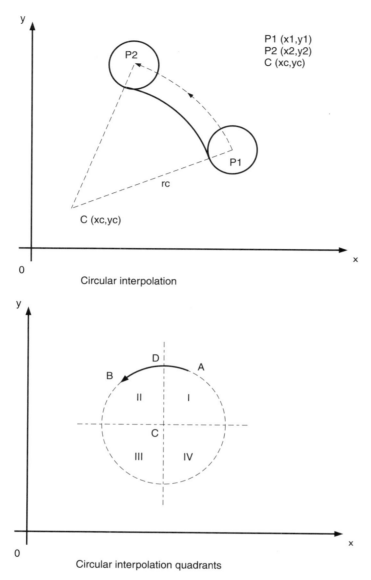

Figure 4.4: Circular interpolation.

The interpolation parameter K is used when one of the axes is Z. Some CNC systems allow circular arcs to be programmed only in one quadrant at a time. Each arc must be programmed in a separate NC block. For example, the circular arc shown in Figure 4.4 must be programmed in two NC blocks:

N010 G90 G03 $X\, x_D$ $Y\, y_D$ $I\,(x_C - x_A)$ $J\,(y_C - y_A)$ F200
N020 G03 $X\, x_B$ $Y\, y_B$ $I\,(x_C - x_D)$ $J\,(y_C - y_D)$

Note that new-generation CNC systems allow programming of more than one arc quadrant in one block.

G81 to G89 – Fixed Cycles

The fixed cycle (or canned cycle) commands set up the CNC unit for different types of automatic operations. Use of fixed cycles permits the reading of X–Y location, a rapid Z plane plunge (programmed with R word), a final Z feed point and a feed rate into the control. The values of the words are stored until they are replaced with new numbers or canceled by G80. By the sole reading of a new X and Y location, the stored sequence of events is performed at the new location. The following NC codes illustrate the use of a canned drill cycle:

```
N050  G81  X125.0  Y237.5  Z-112.5  R2.5  F300
N060        X325.0
N070                    Y137.5
N080  G80
```

In block N050, the cutter is positioned at rapid traverse to 125.0 mm in X and 237.5 mm in Y. The spindle would then speed to 2.5 mm in Z and feed at 300 mm/min to a Z depth of -112.5 mm. The final motion for the block would be rapid retraction of the spindle to the R plane of 2.5 mm in Z. The subsequent blocks N060 and N070 would drill an additional two holes at the indicated new locations. The G80 in block N080 cancels the fixed drill cycle G81.

It must be noted that some machine tool control units may use different formats and codes from the listed standard codes.

An NC part program listing for the sample workpiece profile shown in Figure 4.5 is given as follows:

```
N01G90                         ; Absolute coordinates
N02G71                         ; Metric units (mm)
N03G92X-12.5 Y-12.5 Z50.0      ; Cutter starts from here
                                 with respect to the part zero
N04G00Z2.5 M03 S800            ; Spindle on CW, move rapidly
                                 to 2.5 mm above the part
N05G01Z-7.5 F25.0 M08          ; Coolant on, plunge
                                 with feed in Z (P1)
N06    X162.5 F125             ; Move in x with 125 mm/min
                                 feed to P2
N07    Y0.0                    ; Move to P3 at the same
                                 feed (125 mm/min)
N08G02X220 Y57.5 I57.5 J0      ; CW circular interpolation (P4)
N09G01X232.5                   ; Move one cutter radius in x
N10    Y70.0                   ; Move in y (P6)
N11G03X180 Y122.5 I-52.5 J0    ; CCW relative to the tool
                                 motion in block N10 (P7)
N12G01X107.5                   ; Move left (P8)
N13    Y110                    ; Move to (P9)
N14G02X80.0 Y82.5 I-27.5 J0    ; CW circular interpolation (P10)
N15G01X40.0                    ; Move left (P11)
N16G03X-12.5 Y30.0 I0 J-52.5   ; CCW circular interpolation (P12)
```

Figure 4.5: Sample workpiece for NC programming.

Points	X	Y	Circle center
P0	0	0	
P1	-12.5	-12.5	
P2	-162.5	-12.5	
P3	162.5	0.0	
P4	220	57.5	c(220,0)
P5	232.5	57.5	
P6	232.5	70	
P7	180	122.5	c(180,70)
P8	107.5	122.5	
P9	107.5	110	
P10	80	82.5	c(80,110)
P11	40	82.5	
P12	-12.5	30	c(40,30)
P1	-12.5	-12.5	

```
N17G01Y-12.5          ; Return to starting point (P1)
N18    Z3.8           ; Move up to z = 3.8 mm position
N19G00Z50 M09 M05     ; Move to Z = 50 rapidly,
                        coolant and spindle are off
N20M30                ; Program ends, move the cutter
                        to starting point.
```

4.6 COMPUTER-ASSISTED NC PART PROGRAMMING

It is rather tedious and unproductive to generate NC part programs manually, as demonstrated in the previous section. In computer-assisted NC part programming, the computer automatically generates the codes. The philosophy behind the computer-assisted programming is rather straightforward to describe. If a part geometry can be parametrically defined in a computer code, the tool path can also be generated for given dimensions of a tool according to desired sequences of machining. The programmer must, of course, select tools, feeds, and speeds and generate the tool paths according to the process planning strategy for the machining operation.

Computer-assisted part programming can be done with and without the aid of computer graphics. Both methods require mathematical representation of geometric entities in a computer program. Before introducing the computer-aided NC programming techniques used in industry, the fundamental principles of analytical geometry used in CAD/CAM systems are briefly presented in the following section.

4.7 BASICS OF ANALYTICAL GEOMETRY

The scientific principles or even the technology of computer-aided design and computer-aided manufacturing (CAD/CAM) are too widespread to cover in this text. However, a basic introduction to geometric entities and their manipulation in both CAD and CNC machines is essential for leading readers to more advanced topics related to CAD/CAM. As briefly presented in NC programming with APT and advanced computer graphic tools, the geometry of the workpiece is represented by points, lines, circles, splines, surfaces, and solids. The geometric entities are manipulated by translation, rotation, intersection, and trimming operations. The objects are either represented by their wire frame or solid models. Using shading and hidden line-surface removal techniques, the user can visualize the designed part on the graphics terminal. Because the detailed presentation of CAD techniques and computer graphics are well covered in dedicated texts [67, 68], instead, we will briefly introduce the mathematical formulation of the very basic geometric entities used in both CAD and real-time CNC interpolation algorithms.

Figure 4.6: Vectors and lines.

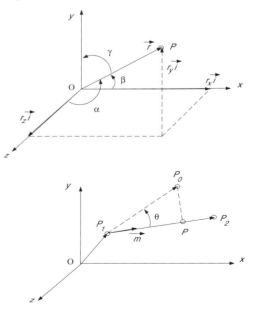

4.7.1 Vectors and Lines

A vector in a Cartesian coordinate system is expressed in Fig. 4.6 as

$$\vec{r} = r_x\vec{i} + r_y\vec{j} + r_y\vec{k}, \qquad (4.1)$$

where $r_x = r\cos\alpha, r_y = r\cos\beta$, and $r_z = r\cos\gamma$ are the directional factors. The magnitude of the vector is $|\vec{r}| = (r_x^2 + r_y^2 + r_y^2)^{1/2}$. The unit vector, which has a magnitude of unity, is $\vec{r}/|\vec{r}| = \cos\alpha\vec{i} + \cos\beta\vec{j} + \cos\gamma\vec{k}$.

A vector between two points ($P_1(x_1, y_1, z_1)$, $P_2(x_2, y_2, z_2)$) can be constructed as

$$\vec{P_1P_2} = (x_2 - x_1)\vec{i} + (y_2 - y_1)\vec{j}$$
$$+ (z_2 - z_1)\vec{k}. \qquad (4.2)$$

A unit vector along $\overrightarrow{P_1 P_2}$ is

$$\vec{m} = \frac{\overrightarrow{P_1 P_2}}{|\overrightarrow{P_1 P_2}|} = \frac{x_2 - x_1}{L}\vec{i} + \frac{y_2 - y_1}{L}\vec{j} + \frac{z_2 - z_1}{L}\vec{k}, \tag{4.3}$$

where $L = |\overrightarrow{P_1 P_2}| = [(x_2 - x_1)^2 + (y_2 - y_1)^2 + (z_2 - z_1)^2]^{1/2}$ is the length of the line connecting the two points.

The products of vectors provide angles, projections, and distances between various geometric entities. Consider two vectors $\vec{A} = \overrightarrow{P_1 P_0} = A_x\vec{i} + A_y\vec{j} + A_y\vec{k}$ and $\vec{B} = \overrightarrow{P_1 P_2} = B_x\vec{i} + B_y\vec{j} + B_y\vec{k}$. The dot product of the two vectors is a scalar:

$$\vec{A} \cdot \vec{B} = A_x B_x + A_y B_y + A_z B_z = |\vec{A}||\vec{B}| \cos\theta, \quad 0 \le \theta \le \pi. \tag{4.4}$$

Note that $\vec{i} \cdot \vec{i} = \vec{j} \cdot \vec{j} = \vec{k} \cdot \vec{k} = 1$ and $\vec{i} \cdot \vec{j} = \vec{j} \cdot \vec{k} = \vec{k} \cdot \vec{i} = 0$. Hence, the angle between the two vectors can be evaluated by $\theta = \arccos[\vec{A} \cdot \vec{B}]/[|\vec{A}||\vec{B}|]$.

The vector product of $(\vec{A}, \vec{B})$ is a vector:

$$\begin{aligned} \vec{A} \times \vec{B} &= (A_y B_z - A_z B_y)\vec{i} + (A_{zy} B_x - A_x B_z)\vec{j} + (A_x B_y - A_y B_x)\vec{k} \\ &= (|\vec{A}||\vec{B}| \sin\theta)\vec{n}, \quad 0 \le \theta \le \pi, \end{aligned} \tag{4.5}$$

where $\vec{n}$ is a unit vector perpendicular to the plane formed by vectors $\vec{A}$, and $\vec{B}$, and θ is the angle between the two vectors. The distance between a point and a line can be evaluated using vector algebra. The unit vector of line $P_1 P_2$ is $\vec{m} = \overrightarrow{P_1 P_2}/|\overrightarrow{P_1 P_2}|$. The distance between point P_0 and line $P_1 P_2$ is (Fig. 6.4)

$$|\overrightarrow{P_0 P}| = |\overrightarrow{P_1 P_0} \times \vec{m}| = |\overrightarrow{P_1 P_0}| \cdot |\vec{m}| \sin\theta.$$

4.7.2 Translation and Rotation of Objects

Translating an object from one location to another requires that every point and entity describing it experience the same amount of displacement. If the displacement vector is $\vec{d}$, and the object to be translated is $\vec{P}$, than the translated position of the object becomes

$$\vec{P}^* = \vec{P} + \vec{d}. \tag{4.6}$$

The objects can be scaled by simply multiplying their coordinates,

$$\vec{P}^* = [S]\vec{P}, \tag{4.7}$$

where the scale matrix $[S]$ is

$$[S] = \begin{bmatrix} s_x & 0 & 0 \\ 0 & s_y & 0 \\ 0 & 0 & s_z \end{bmatrix}.$$

An object can be rotated about the z axis, and the relevant equation is

$$\vec{P}^* = [R_z]\vec{P}, \tag{4.8}$$

where the rotation matrix $[R_z]$ is

$$[R_z] = \begin{bmatrix} \cos\theta & -\sin\theta & 0 \\ \sin\theta & \cos\theta & 0 \\ 0 & 0 & 1 \end{bmatrix},$$

where θ is the counterclockwise rotation angle. Similarly, the rotation matrices for the x and y axes are given by

$$[R_x] = \begin{bmatrix} 1 & 0 & 0 \\ 0 & \cos\theta & -\sin\theta \\ 0 & \sin\theta & \cos\theta \end{bmatrix},$$

$$[R_z] = \begin{bmatrix} \cos\theta & 0 & \sin\theta \\ 0 & 1 & 0 \\ -\sin\theta & 0 & \cos\theta \end{bmatrix}.$$

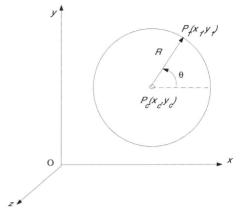

Figure 4.7: Mathematical representation of circles.

Rotation with respect to an arbitrary axis can be derived using vector algebra [68].

4.7.3 Circles

A point $P(x, y, z)$ on a circle (Fig. 4.7) can be defined by the radius and angular distance,

$$P(x, y, z) = \begin{Bmatrix} x \\ y \\ z \end{Bmatrix} = \begin{Bmatrix} x_c + R\cos\theta \\ y_c + R\sin\theta \\ z_c \end{Bmatrix}, \quad 0 \leq \theta \leq 2\pi. \tag{4.9}$$

The following equation can be used to identify the center (x_c, y_c) and radius r of a circle passing through three points:

$$r^2 = (x - x_c)^2 + (y - y_c)^2. \tag{4.10}$$

Alternatively, the three points can be used to evaluate the coefficients (c_1, c_2, c_3) of a nonparametric equation defining the circle in the (x, y) plane:

$$y = c_1 + c_2 x + c_3 x^2.$$

4.7.4 Cubic Splines

Parts such as dies, molds, and gas turbine blades have sculptured surfaces that are designed by fitting smooth curves to a set of design points. Although there are various spline fitting techniques available in CAD systems, they can all be derived by understanding the algorithm for a cubic spline fit. A series of sequential design points $P_0, P_1, \ldots, P_{i-1}, P_i, P_{i+1}, \ldots, P_n$ are shown in Figure 4.8. The points must be connected by series of cubic spline segments. Consider an arbitrary point $P_i(u)$ on the cubic spline segment S_i:

$$P_i(u) = A_i u^3 + B_i u^2 + C_i u + D_i, \quad 0 \leq u \leq 1. \tag{4.11}$$

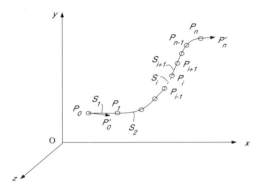

Figure 4.8: Cubic spline curve fit.

Note that the end points of the spline segment are $P_i(u = 0) \equiv P_{i-1}, P_i(u = 1) \equiv P_i$. By increasing the parameter u from zero toward unity with small increments, the spline segment can be generated from a series of points provided that the spline parameters A_i, B_i, C_i, and D_i are known. The derivative of Eq. (4.11) gives the tangent vectors

$$P'_i(u) = 3A_i u^2 + 2B_i u + C_i. \tag{4.12}$$

Using the boundary conditions at both ends of the spline segment, the parameters A_i, B_i, C_i, and D_i are evaluated. The resulting expression for the cubic spline and its tangent and curvature expressions are

$$
\begin{aligned}
P_i(u) &= [2(P_{i-1} - P_i) + P'_{i-1} + P'_i]u^3 \\
&\quad + [3(P_i - P_{i-1}) - 2P'_{i-1} - P'_i]u^2 \\
&\quad + P'_{i-1}u + P_{i-1}, \\
P'_i(u) &= [6(P_{i-1} - P_i) + 3(P'_{i-1} + P'_i)]u^2 \\
&\quad + [6(P_i - P_{i-1}) - 4P'_{i-1} - 2P'_i]u + P'_{i-1}, \\
P''_i(u) &= [12(P_{i-1} - P_i) + 6(P'_{i-1} + P'_i)]u \\
&\quad + [6(P_i - P_{i-1}) - 4P'_{i-1} - 2P'_i].
\end{aligned} \tag{4.13}
$$

Similar expressions can be written for the spline segment S_{i+1} bounded by points P_i and P_{i+1}. Like an elastic beam, curvature continuity is imposed at the knots where the spline segments S_i and S_{i+1} meet [i.e., $P''_i(u = 1) = P''_{i+1}(u = 0)$]. The resulting expression provides a relationship between the unknown tangent vectors and known points:

$$P'_{i-1} + 4P'_i + P'_{i+1} = 3P_{i+1} - 3P_{i-1}. \tag{4.14}$$

In order to solve all tangent vectors at every knot, the tangents at the two end points (P_0, P_n) must be provided. One can either impose a tangent depending on the design criteria or adjacent geometric entities, or one can approximate them. For example, one approach is to use the lines connecting the first and last knots with their nearest neighbors, and use their slope:

$$P'_0 = \frac{\overrightarrow{P_0 P_1}}{|\overrightarrow{P_0 P_1}|}, \qquad P'_n = \frac{\overrightarrow{P_{n-1} P_n}}{|\overrightarrow{P_n P_{n-1}}|}. \tag{4.15}$$

Alternatively, the two ends can be relaxed like a beam with moment-free ends (i.e., $P_0'' = P_n'' = 0$), which leads to two more required expressions:

$$
\begin{aligned}
2P_0' + P_1' &= 3P_1 - 3P_0, \\
2P_n' + P_{n-1}' &= 3P_n - 3P_{n-1}.
\end{aligned}
\tag{4.16}
$$

Equations (4.14) and (4.16) can be combined in the following matrix:

$$
\begin{bmatrix}
2 & 1 & 0 & 0 & . & . & . & 0 & 0 & 0 \\
1 & 4 & 1 & 0 & . & . & . & 0 & 0 & 0 \\
. & . & . & . & & & & . & . & . \\
. & . & . & . & & & & . & . & . \\
. & . & . & . & & & & . & . & . \\
0 & 0 & 0 & . & . & . & 1 & 4 & 1 & 0 \\
0 & 0 & 0 & . & . & . & 0 & 1 & 4 & 1 \\
0 & 0 & 0 & . & . & . & 0 & 0 & 1 & 2
\end{bmatrix}
\begin{bmatrix}
P_0' \\ P_1' \\ P_2' \\ . \\ . \\ . \\ P_{n-2}' \\ P_{n-1}' \\ P_n'
\end{bmatrix}
= 3
\begin{bmatrix}
P_1 - P_0 \\ P_2 - P_0 \\ P_3 - P_1 \\ . \\ . \\ . \\ P_{n-2} - P_{n-4} \\ P_{n-1} - P_{n-3} \\ P_n - P_{n-1}
\end{bmatrix}.
\tag{4.17}
$$

Hence, the tangents at all knots can be evaluated by solving matrix Eq. 4.17. Substituting the tangents of each spline segment into the cubic Eq. 4.13 yields to a spline fit equation:

$$
\begin{aligned}
P_i(u) &= [\, A_i \ \ B_i \ \ C_i \ \ D_i \,][\, u^3 \ \ u^2 \ \ u \ \ 1 \,]^T \\
&= [-2u^3 + 3u^2]P_{i+1} + [2u^3 - 3u^2 + 1]P_i \\
&\quad + (u^3 - u^2)P_{i+1}' + (u^3 - 2u^2 + u)P_i', \qquad 0 \le u \le 1.
\end{aligned}
\tag{4.18}
$$

There are better curve-fitting techniques for smoother design and NC tool path generation. Some of the best-known methods include Bezier curves [67] and bicubic splines. The splines are used to model the geometry and NC tool paths of parts with sculptured surfaces. Higher order splines are also used to generate the splined tool paths in real-time CNC applications.

4.8 APT PART PROGRAMMING LANGUAGE

Owing to difficulties in manual NC programming, a computer language APT (Automatically Programmed Tools) was developed at MIT (1956) to automate part programming. Like any other high-level language, APT allows arithmetic operations, subroutines, macros, looping logic, etc. However, the main difference is that it allows parametric representations of geometric entities ranging from a single point in space to very complex three-dimensional sculptured surfaces. The tool path is generated by commanding the given tool geometry to travel in a selected tool path on the defined part geometry. The computed tool path coordinates and cutting conditions are then stored in a computer file with a general standard format, which is also automatically converted to the specific manual code of the selected CNC machine tool.

APT was the most common and standard code used in the industry. APT has been replaced by graphics-based, user-friendly interactive CAD/CAM software systems in industry. However, since CAD/CAM systems use APT standards in processing NC tool path data, it is useful to understand at least the very basic structure of the APT programming system.

4.8.1 Geometric Statements

The part geometry is defined using points, lines, arcs, space curves, and surfaces. Each geometric entity is mathematically represented and stored as variables by APT. Basics of analytical geometry are presented in the previous section. The user must define each geometric entity according to the format required by APT. The formats and the use of geometric statements can be found in the APT dictionary generated by CAM-I (Computer Aided Manufacturing-International) [69]. Each geometry is defined according to the following format:

(Geometry's symbol) = **GEOMETRY**/(Dimensions and the parameters of the geometry).

Some of the fundamental geometric statements are illustrated in the following. Note that the lowercase letters represent labels or dimensions to be entered by the programmer; capital letters and operator symbols (, / = $) must be used as they are shown.

Point

The following statements can be used in defining a point.

- **Pname = POINT/ x, y, z**
  Here x, y, and z are the coordinates of the point. Zero value is assumed when the z dimension is omitted.
  Example: P1 = POINT/2.0,3.2,2.0

- **Pname = POINT/INTOF, line, line**
  The point is defined by the intersection of two lines, which are defined by earlier statements.
  Example: P1 = POINT/INTOF, L1, L2

- **Pname = POINT/{XLARGE, XSMALL, YLARGE, YSMALL}, INTOF, Line, Circle**
  If there are multiple possible points, only one of the *XLARGE, XSMALL, YLARGE,* and *YSMALL* identifiers is used to distinguish one point from the others.
  Example (Figure 4.9): P1 = POINT/XSMALL, INTOF, L1, C1
  There are two possible intersection points between line L1 and circle C1, but point P1 has smaller x dimension than point P2.

- **Pname = POINT/{XLARGE, XSMALL, YLARGE, YSMALL}, INTOF, Circle, Circle**
  Example (Figure 4.10): P1 = POINT/Y LARGE, INTOF, C1, C2

- **Pname = POINT/Circle, ATANGL, angle**
  Example : (Figure 4.11): P1 = POINT/C1, ATANGL, 60.0

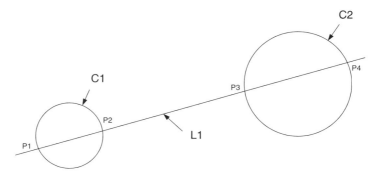

Figure 4.9: Point definition with statement: P1 = POINT/XSMALL, INTOF, L1, C1.

Note that the angle is measured positive in the CCW direction using the x axis as a reference.

- **Pname = POINT/CENTER, circle**
  The point is the center of a circle.

- **Pname = POINT/point, THETAR, angle, radius**
  Example (Figure 4.12): P2 = POINT/P1, THETAR, 30, 22.5
  Note that the angle is measured positive in the CCW direction using the x axis as a reference.

Lines

The following set of geometric commands are provided as line construction examples in APT.

- **Lx = LINE/P1, P2**
  Line between points P1 and P2.

- **Lx = LINE/point, {RIGHT,LEFT}, TANTO, Circle**
  Line passing from a point and tangent to the right or left of a circle. RIGHT or LEFT descriptive words are used by looking from the point (first geometry used in the LINE command) toward the circle (second geometry used in the LINE command). Figure 4.13 shows how two lines are constructed using the command.

- **Lx = LINE/{RIGHT,LEFT}, TANTO, circle1, {RIGHT,LEFT}, TANTO, circle2**
  Figure 4.13 illustrates the definition of one line out of the four possible tangent lines to two circles.

- **Lx = LINE/point, PARLEL, line**
  Line passing from a point and parallel to another line (see Fig. 4.14).

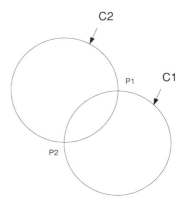

Figure 4.10: Point definition with statement: P1 = POINT/ YLARGE, INTOF, C1, C2.

C1

P1

60°

Figure 4.11: Point definition with statement: P1 = POINT/C1, ATANGL, 60.0.

- **Lx = LINE/{XLARGE, XSMALL, YLARGE, YSMALL}, TANTO, circle, ATANGL, angle**
  Line tangent to a circle that has an angle with the x axis. One of the descriptive words XLARGE, XSMALL, YLARGE, YSMALL is used to select the line out of more possible lines (see Figure 4.15).

Circles

The use of the following circle definition syntaxes is shown on the indicated figures.

1. Circlex = CIRCLE/(x,y,z), Radius
   = CIRCLE/CENTER, point, RADIUS, radius
   = CIRCLE/CENTER, point, TANTO, line

   Example (Figure 4.16): C1=CIRCLE/CENTER, (POINT/1,2), TANTO, (LINE/P3,P4)
2. Circlex=CIRCLE/CENTER, center point, point on the circle
3. Circlex=CIRCLE/point1,point2,point3 (three points on the circle)
4. Circlex=CIRCLE/TANTO,line,{XSMALL..YLARGE}, point, RADIUS, radius
   Example (Figure 4.17): C2=CIRCLE/TANTO,L1,XSMALL,P1, RADIUS, 20
5. Circlex=CIRCLE/{XSMALL..YLARGE}, line,{XSMALL..YLARGE}, {IN, OUT}, Circle, RADIUS, radius
   Example: C1=CIRCLE/YLARGE, LG, XSMALL, OUT, CG, RADIUS, 10
6. Circlex = CIRCLE/CENTER, Point, {SMALL,LARGE}, TANTO, Circle
   Example (Figure 4.18): Both circles C0 and C2 are tangent to the circle C1, but C0 is *smaller* than C2.

   C0 = CIRCLE/CENTER, P1, SMALL, TANTO, C1
   C2 = CIRCLE/CENTER, P1, LARGE, TANTO, C1

Figure 4.12: Point definition with statement: P2 = POINT/P1, THETAR, 30, 22.5.

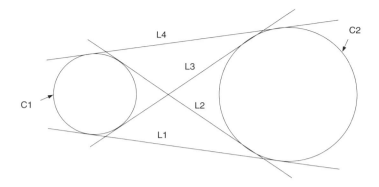

L1 = LINE / RIGHT, TANTO, C1, RIGHT, TANTO, C2
L2 = LINE / LEFT, TANTO, C1, RIGHT, TANTO, C2

Figure 4.13: Line definition with statement: L1 = LINE/{RIGHT, LEFT}, TANTO, circle1, {RIGHT,LEFT}, TANTO, circle2.

7. Circlex=CIRCLE/{XLARGE..YSMALL},line1, {XSMALL..YLARGE}, line2, RADIUS, radius

Example (Figure 4.19): Out of four possible circles, only C1 has a larger y dimension with respect to L2 and smaller x dimension with respect to line L1.

C1=CIRCLE/YLARGE,L2,XSMALL, L2, RADIUS, 10

Planes

Planes are generally used to define the planes of motion for the tool movements. Its surface normal vector and its amplitude define a plane:

Planex= PLANE/a,b,c,d

where a, b, and c are the cosines of the angles between the surface normal vector and the plane, and d is the length of the vector.

Example: PL1=PLANE/0, 0, 1, 7.5

Plane PL1 is parallel to the xy plane, and the height is 7.5 along the z axis.

Complex Curves

APT allows the use of parabolas, hyperbolas, ellipses, and parametric curves. Splines are the most common parametric curves that allow flexible and smooth curve fit passing from a given set of points. For example, a spline curve fit passing from n points is given by (see Figure 4.20)

$$\text{Spline1} = \text{SPLINE/Point}_1, \text{Point}_2, \dots, \text{Point}_n$$

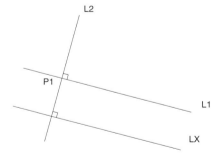

L1 = LINE/P1, PARLEL, LX
L2 = LINE/P1, PERPTO, LX

Figure 4.14: Line passing from a point and parallel to another line: Lx = LINE/ point, PARLEL, line.

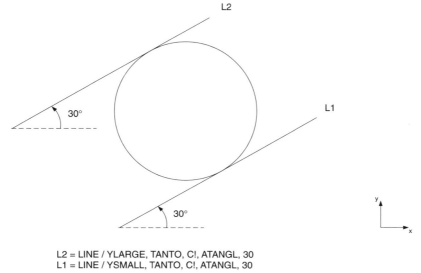

L2 = LINE / YLARGE, TANTO, C!, ATANGL, 30
L1 = LINE / YSMALL, TANTO, C!, ATANGL, 30

Figure 4.15: Lx = LINE/XLARGE, XSMALL, YLARGE, YSMALL, TANTO, circle, ATANGL, angle.

Sculptured Surfaces

In order to machine molds, dies, aircraft, and complex parts, their sculptured surfaces must be represented smoothly by mathematical models. There are a number of available parametric surface fitting algorithms. Most methods use bicubic, Coon's surfaces, and Bezier surfaces [67]. APT and commercial CAD/CAM packages employ both surface fitting techniques as well as parametric cubic fitting methods. Some of the basic surface types used in both APT and CAD/CAM systems are introduced as follows.

Figure 4.16: C1 = CIRCLE/CENTER, (POINT/1.0,2.0), TANTO, (LINE/P3,P4).

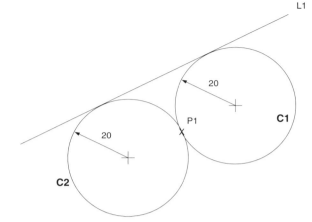

Figure 4.17: C2 = CIRCLE/TANTO, L1, XSMALL, P1, RADIUS, 20.

Rule Surfaces. An aircraft wing is a typical example for a rule surface. First the root and the tip airfoils are constructed using spline curve fits, and a ruled surface is constructed between the two airfoils. Connecting each point on the root airfoil with the corresponding one on the tip generates the surface. The number of points on the root and tip airfoils must be identical and must have the same order of definition:

Surface=SSURF/SRUL,XYZ,SPLINE,spline,SPLINE,spline

Example: An aircraft wing surface fit by the following ruled surface command is shown in Figure 4.20:

S1   = SPLINE/$P_1, P_2, \ldots, P_n$
S2   = SPLINE/$Q_1, Q_2, \ldots, Q_n$
WING = SSURF/SRUL, XYZ, SPLINE, S1, SPLINE, S2

Figure 4.18: Both circles C0 and C2 are tangent to the circle C1, but C0 is smaller than C2.

C0 = CIRCLE/CENTER, P1, SMALL, TANTO, C1
C1 = CIRCLE/CENTER, P1, LARGE, TANTO, C1

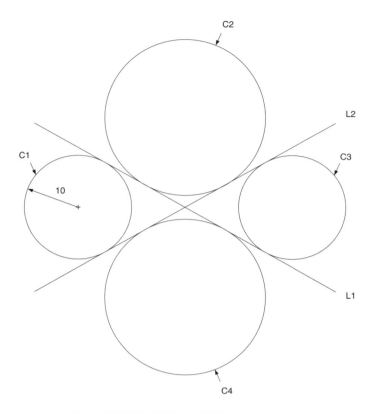

C1 = CIRCLE/LARGE, L2, XSMALL, L1, RADIUS, 10

Figure 4.19: Out of four possible circles, only C1 has larger y dimension with respect to L2 and smaller x dimension with respect to line L1.

Twisted Surfaces. An aircraft propeller or a gas turbine blade is a perfect example for twisted surface fitting in APT.

Example:

S1 = SPLINE/$P_1, P_2, \ldots, P_n$
S2 = SPLINE/$P_1, P_2, \ldots, P_n$
⋮
Sn = SPLINE/$P_1, P_2, \ldots, P_n$
BLADE = SSURF/SMESH, XYZ, SPLINE, S1, SPLINE, S2,..., SPLINE, Sn

4.8.2 Tool Motion Statements

There are two types of tool motion commands in APT: point to point (PTP) and continuous tool motion commands.

Point-to-Point Tool Motion Commands

If the tool is required to move to a specified point from its current position, PTP commands are used. There are two types of PTP motion commands,

GOTO/point, feedrate

S1=SPLINE/P1,P2,P3,....,Pn

S2=SPLINE/Q1,Q2,Q3,....,Qn
WING=SSURF/SRUL,XYZ,SPLINE,S1,SPLINE,S2.

Figure 4.20: APT commands for spline (S1,S2) and ruled surface (WING) fit from a given set of points.

and

GODLTA/point, feedrate

Both commands take the tool to the point indicated to the right of the slash with an optional feed rate provided. The point coordinates are assumed to be given in absolute coordinates (G90) by a GOTO command and incremental (G91) coordinates by a GODLTA command. If only one coordinate is given in the GODLTA/ command, incremental z axis motion is assumed.

Example: GOTO/10,2,1

The tool is commanded to move to the point that has absolute coordinates of $x = 10$ mm, $y = 2$ mm, and $z = 1$ mm.

Example: GODLTA/10,2,1

The tool is commanded to move incrementally 10 mm, 2 mm, and 1 mm in the x, y, and z axis respectively.

Example: GODLTA/-10

The tool is commanded to move 10 mm in the negative z axis direction.

Continuous Tool Motion Commands

The tool is commanded to move along a controlled path defined by the *part, drive,* and *check* surfaces. If we consider a three-axis machining process with a vertical CNC milling machine tool, the tool's end stays on the part surface, the

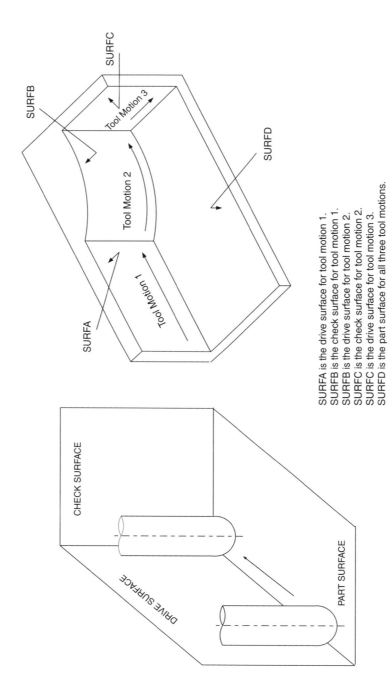

SURFA is the drive surface for tool motion 1.
SURFB is the check surface for tool motion 1.
SURFB is the drive surface for tool motion 2.
SURFC is the check surface for tool motion 2.
SURFC is the drive surface for tool motion 3.
SURFD is the part surface for all three tool motions.

Figure 4.21: Feed motion directions in APT.

tool's axis stays parallel to the drive surface, and the tool motion is bounded by the check surface (see Fig. 4.21.) If the part surface is a three-dimensional sculptured surface (i.e., an aircraft wing or a turbine blade surface), a three-axis machine tool is required. If two of the surfaces are sculptured then five or more simultaneously controlled axes are required on the CNC machining center.

Before the tool can be commanded to move along a given profile, it must be brought from a specific point. Usually this point corresponds to tool-change location coordinates with respect to the part program zero. This is achieved by a

FROM/point

command, followed by

GO/{TO,ON,PAST}, drive surface, {TO, ON, PAST}, part surface, {TO, ON, PAST}, check surface

where one of the three modifiers (TO, ON, PAST) is selected to define the position of the tool with respect to the surface. In direction vector (INDIRV/vector) or in direction point (INDIRP/point) commands may be used ahead of GO/... statements to direct the tool motion in a desired direction. These three statements are often called *startup* commands in APT vocabulary. The startup commands are typically followed by continuous motion statements:

$$\left\{\begin{array}{l} \text{GOFWD} \\ \text{GOBACK} \\ \text{GOLFT} \\ \text{GORGT} \end{array}\right\} \text{/drive surface,} \left\{\begin{array}{l} \text{TO} \\ \text{ON} \\ \text{PAST} \\ \text{TANTO} \end{array}\right\}, \text{check surface}$$

where GOFWD, GOBACK, GOLFT, and GORGT motion statements correspond to continuous tool motion movement in the *forward, backward, left*, or *right* direction relative to the preceding motion (see Figures 4.22 and 4.23).

Example: A typical motion command sequence might be:

```
GO        / TO, L1, ON, PL1, TO, C1
GOFWD     / C1, TO, L1
GOLFT     / L1, ON, L1
```

4.8.3 Cutter Location File and Postprocessing

APT processor or CAD/CAM systems produce two cutter location (CL) files. The first is in readable ASCII, and the second is in binary format. The ASCII printout can be obtained by inserting a CLPRNT/ON command before the motion statements. The center coordinates of the tool are printed for the programmer to debug the APT program. The binary file has a standard format. Each record in the file may have two or more fields depending on the type of command. The length of each field is again several bytes long and is determined by the command's function. For example,

Program	Class	Code	Field
PARTNO / A0010	2000	1045	A0010
TOOL / 14	2000	1025	10
COOLNT / ON	2000	1030	71
FEDRAT / 30.0	2000	1009	30
FROM / 0,0, 3	5000	3	0.000 0.000 0.000
GODLTA / 0.3	5000	4	0.300
GOTO / 5.0, 3.7	5000	5	5.000 3.700
FINI	2000	1	

Class word defines whether the statement produces a miscellaneous function (M function), a point to point motion, or a continuous tool motion. The second word gives the specific code for the statement. The third field, which may be several words long, gives the coordinates of the motion or logic value for the miscellaneous functions. For example, PARTNO/ , TOOL/ , COOLNT/ FINI are called postprocessor commands, and all are classified under 2000. COOLNT has a code value of 1030 and 71 means ON. GOTO/ and GODLTA/ commands belong to the point motion class and are grouped under 5000. GOTO has a code of 5, followed by three double precision floating point long field words that give the x, y, and z coordinates of the target point.

The CL file has an internationally standard format. Each CNC machine tool builder provides a postprocessor for its specific CNC machine tool. After generating the CL file, the programmer *postprocesses* the CL file and obtains the final manual equivalent NC code automatically. Note that if any dimensions

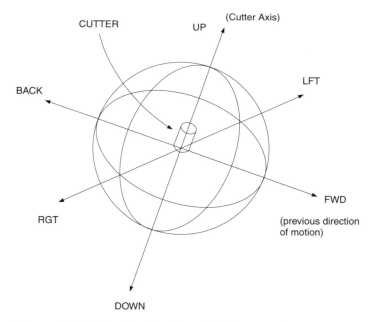

Figure 4.22: Feed motion directions in APT.

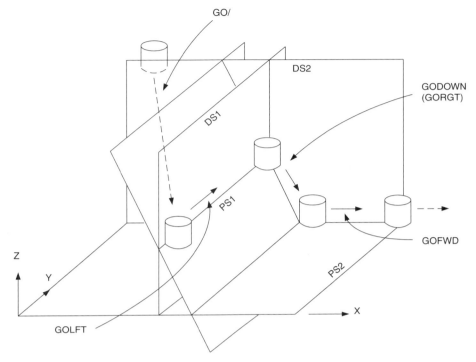

Figure 4.23: Sample motions for different GO motion commands in APT.

changed in the workpiece, the geometric statements can be updated, processed, and postprocessed without having to program the part again.

APT Programming Example I. The manual programming of the part shown in Figure 4.5 was illustrated in the previous section. The APT program for the same part is given in the following.

```
PARTNO MECH
$$
$$ AUTHOR : Y.Altintas, Date: Jan. 25, 1988
$$ Purpose : Demonstration program
$$
CLPRINT $$ Print the CL file
UNITS/mm $$ Metric units
INTOL/0.01 $$ Path following tolerance
OUTTOL/0.01
$$
CUTTER/10,0 $$ 10 mm dia, flat end mill.
$$ Workpiece geometry
P0 = POINT/0.,0.,0. $$ Program zero
SP = POINT/-12.5.,-12.5.,30. $$ Machine zero
P2 = POINT/150.,0.,0.
P3 = POINT/220.,70.,0.
```

P4 = POINT/180.,110.,0.
P5 = POINT/120.,110.,0.
P6 = POINT/80.,70.,0.
P7 = POINT/40.,70.,0.
P8 = POINT/0.,30.,0.
$$ Lines
L1 = LINE/P0,P2
L2 = LINE/P4,P5
L3 = LINE/P6,P7
L4 = LINE/P8,P0
$$ Arcs
C1 = CIRCLE/CENTER,(POINT/220.,0.),RADIUS,70.
C2 = CIRCLE/CENTER,(POINT/180.,70.),RADIUS,40.
C3 = CIRCLE/CENTER,(POINT/80.,110.),RADIUS,40.
C4 = CIRCLE/CENTER,(POINT/40.,30.),RADIUS,40.
$$ PLANES
PL1 = PLANE/0,0,1,30. $$ Z=30mm PLANE
PLB = PLANE/0,0,1,-10. $$ Z=-10mm PLANE
PLC = PLANE/0,0,1,5. $$ Z=5mm PLANE
PLXZ = PLANE/0,1,0,0 $$ Y=0 PLANE (XZ)
PLYZ = PLANE/1,0,0,0 $$ X=0 PLANE (YZ)
$$ MOTION COMMANDS
FROM/SP $$ STARTING POINT
$$ Continuous path initialization
GO/TO,L1,TO,PLC,TO,L3
FEDRAT/50 $$ 50mm/min plunge feed
GO/TO,L1,TO,PLB,TO,L3 $$ PLUNGE IN Z
PSIS/PLB $$ PART SURFACE IS PLB
INDIRV/1,0,0
$$ Continuous path contouring starts
GOFWD/L1,PAST,(L11=LINE/P2,PARLEL,L3)
GOLFT/L11,ON,L1
GOFWD/C1,PAST,(L12=LINE/P3,PARLEL,L3)
GOLFT/L12,ON,(L13=LINE/P6,PARLEL,L1)
GOFWD/C2,PAST,L2,ON,(L14=LINE/P4,PARLEL,L3)
GOFWD/L2,PAST,(L15=LINE/P5,PARLEL,L3)
GOLFT/L15,PAST,C3,ON,L2
GOFWD/C3,ON,(L16=LINE/P6,PARLEL,L3)
GORGT/L22,ON,(L17=LINE/P7,PARLEL,L3),TANTO,C4
GOFWD/C4,PAST,L3,ON,(L18=LINE/P8,PARLEL,L1)
GOFWD/L3,PAST,L1 $$ The last contouring motion
GODLTA/12 $$ Go up to the clearance plane
RAPID
GOTO/ SP $$ Go back to the machine zero
FINI

APT Programming Example II. The following are geometric and motion commands for ball end milling of a wing surface [70].

```
$$ SYMMETRIC WING, DESIGNED AND MANUFACTURED BY
$$ YUSUF ALTINTAS 25-SEPT.-79 U.N.B. PREDERICTION/CANADA
$$ GEOMETRIC DEFINITION OF AIRCRAFT WING SURFACE
$$ X: IS THE COORDINATE ALONG THE WING SPAN
$$ Y: IS THE COORDINATE ALONG THE AIRFOIL CHORD
$$ Z: IS THE COORDINATE OF AIRFOIL'S HEIGHT
$$ Wing dimensions are reduced by 40,000 times.
PARTNO AIRCRAFT WING
UNITS/INCH $$ Use imperial units
INTOL/0.001 $$ Maximum deviation from true
OUTTOL/0.001 $$ sculptured surface points.
MACHIN/BPT,4,OPTION,11,1 $$ Bridgeport CNC postprocessor
$$ Define aerodynamic points at the root airfoil
P1 =POINT/ 0.0, 0.00000, 0.00000
P2 =POINT/ 0.0, 0.01562, 0.03078
P3 =POINT/ 0.00000, 0.02344, 0.03731
P4 =POINT/ 0.00000, 0.03906, 0.04747
P5 =POINT/ 0.00000, 0.07312, 0.06569
P6 =POINT/ 0.00000, 0.15625, 0.09141
P7 =POINT/ 0.00000, 0.23437, 0.11069
P8 =POINT/ 0.00000, 0.31250, 0.12622
P9 =POINT/ 0.00000, 0.46875, 0.14997
P10 =POINT/ 0.00000, 0.62500, 0.16694
P11 =POINT/ 0.00000, 0.78125, 0.17850
P12 =POINT/ 0.00000, 0.93750, 0.18531
P13 =POINT/ 0.00000, 1.09375, 0.18531
P14 =POINT/ 0.00000, 1.25000, 0.18500
P15 =POINT/ 0.00000, 1.40625, 0.17825
P16 =POINT/ 0.00000, 1.56250, 0.16781
P17 =POINT/ 0.00000, 1.71875, 0.15422
P18 =POINT/ 0.00000, 1.87500, 0.13813
P19 =POINT/ 0.00000, 2.03125, 0.12000
P20 =POINT/ 0.00000, 2.18750, 0.10031
P21 =POINT/ 0.00000, 2.34375, 0.07987
P22 =POINT/ 0.00000, 2.50000, 0.05944
P23 =POINT/ 0.00000, 2.65625, 0.03981
P24 =POINT/ 0.00000, 2.81250, 0.02209
P25 =POINT/ 0.00000, 2.96875, 0.00781
P26 =POINT/ 0.00000, 3.12500, 0.00000

$$ Points on the tip airfoil
Q1 =POINT/ 10.00000, 0.00000, 0.00000
```

```
Q2 =POINT/ 10.00000, 0.00450, 0.00885
Q3 =POINT/ 10.00000, 0.00675, 0.01075
Q4 =POINT/ 10.00000, 0.01125, 0.01366
Q5 =POINT/ 10.00000, 0.02250, 0.01892
Q6 =POINT/ 10.00000, 0.04500, 0.02632
Q7 =POINT/ 10.00000, 0.06750, 0.03188
Q8 =POINT/ 10.00000, 0.09000, 0.03635
Q9 =POINT/ 10.00000, 0.13500, 0.04319
Q10 =POINT/ 10.00000, 0.18000, 0.04808
Q11 =POINT/ 10.00000, 0.22500, 0.05141
Q12 =POINT/ 10.00000, 0.27000, 0.05337
Q13 =POINT/ 10.00000, 0.31500, 0.05400
Q14 =POINT/ 10.00000, 0.36000, 0.05320
Q15 =POINT/ 10.00000, 0.40500, 0.05134
Q16 =POINT/ 10.00000, 0.45000, 0.04833
Q17 =POINT/ 10.00000, 0.49500, 0.04442
Q18 =POINT/ 10.00000, 0.54000, 0.03970
Q19 =POINT/ 10.00000, 0.58500, 0.03456
Q20 =POINT/ 10.00000, 0.63000, 0.02889
Q21 =POINT/ 10.00000, 0.67500, 0.02300
Q22 =POINT/ 10.00000, 0.72000, 0.01712
Q23 =POINT/ 10.00000, 0.76500, 0.01147
Q24 =POINT/ 10.00000, 0.81000, 0.00636
Q25 =POINT/ 10.00000, 0.85500, 0.00225
Q26 =POINT/ 10.00000, 0.90000, 0.00000
$$ Definition of sculptured wing surface
SW=SSURF/SMESH,XYZ
,SPLINE,P1,P2,P3,P4,P5,P6,P7,P8,P9,P10,P11,P12
,P13,P14,P15,P16,P17,P18,P19,P20,P21,P22,P23
,P24,P25,P26
,SPLINE,Q1,Q2,Q3,Q4,Q5,Q6,Q7,Q8,Q9,Q10,Q11,Q12
,Q13,Q14,Q15,Q16,Q17,Q18,Q19,Q20,Q21,Q22,Q23
,Q24,Q25,Q26
$$ Some geometric definitions to assist tool motion commands
MY=MATRIX/MIRROR,ZXPLAN
TRACUT/MY
XYPL=PLANE/0,0,1,0
SP=POINT/2.0,-2.0,3.0
$$ Cutter motion statements
CUTTER/0.3125,0.15625 $$ Ball end mill with 0.15625″ ball radius
SPINDL/700,CCLW,MEDIUM
FROM/SP
INDIRV/(VECTOR/0,0,-1)
FEDRAT/40
GO/XYPL
```

```
OY=LINE/0,0,0,1
OX=LINE 0,0,1,0
OYY=LINE/-0.15625,0.0,-0.15625,1.0
GO/OYY,SW
INDIRV/0,1,1
GOFWD/OYY,TO,XYPL
R=0.15625
DX=0.04
IU=0 $$ IU IS COUNTER FOR BLOCKING
IX=0
ID1) IX=IX+1
IU=IU+1
IF(IU-10)BL1,BL1,BL2
BL2) INSERT/D
IU=0
BL1) IF(IX-250)ID2,ID2,ID3
ID2) INDIRV/1,0,0
XVL=R+IX*DX
XDRVA=LINE/XVL,0,XVL,1
GOFWD/XYPL,TO,XDRVA
INDIRV/0,-1,1
GOFWD/XDRVA,TO,XYPL
IX=IX+1
XVL=R+IX*DX
XDRVB=LINE/XVL,0,XVL,1
INDIRV/1,0,0
GOFWD/XYPL,TO,XDRVB
INDIRV/0,1,1
GOFWD/XDRVB,TO,XYPL
JUMPTO/ID1
ID3) INDIRV/1,0,0
GOFWD/XYPL,PAST,XDRVB
$$ Side cleaning
GODLTA/-0.1
FEDRAT/45
AUTOFS
INDIRV/0,-1,0
GOFWD/XDRVB,PAST,DX
INDIRV/-1,0,0
GOFWD/DX,PAST,DY
GODLTA/-0.05
AUTOPS
INDIRV/0,1,0
GOFWD/DY,PAST,(LINE/P26,Q26)
GODLTA/0.05
```

```
AUTOPS
INDIRV/1,0,0
GOFWD/(LINE/P26,Q26),PAST,XDRVB
INDIRV/0,-1,0
GOFWD/XDRVB,PAST,(LINE/0,-0.3,1,-0.3)
INDIRV/-1,0,0
GOFWD/(LINE/0,-0.3,1,-0.3),PAST,DY
GODLTA/1.0
SPINDL/OFF
RAPID
GOTO/SP
TRACUT/NOMORE
STOP
FINI
```

4.9 NC PART PROGRAMMING WITH CAD SYSTEMS

CAD systems have already replaced APT. Essentially, most CAD systems have started with the APT processor subroutines but have added interactive graphical displays. Each geometric statement can be made interactively using a mouse. The logic of geometry generation is similar to that of the APT commands.

Model within the stock

Tool Path

Solid model after machining

Photograph of machined part

Figure 4.24: Sample tool paths and three-axis ball end milling of a sculptured surface.

CAD has the advantage of allowing visual inspection of the part geometry on the computer workstation. Most current CAD systems allow three-dimensional construction of the part geometry while simultaneously providing top, side, front, and isometric views. They incorporate solid modeling technology that helps the designer to visualize the actual shape of the part more easily.

Once the geometry is designed on the CAD system, the process planner can pull the part to a computer workstation for a tool path generation. The tool paths are also interactively generated on the graphics station. The process planner defines the tool, feed rate, spindle speed, coolant, and program zero in almost identical fashion to APT. The tool path can be seen on the screen for visual interference checking. Once the tool path is accepted, a CL file is generated by the CAD system. The CL file has an APT standard format and thus postprocessors developed for APT can be used to generate the final manual equivalent NC code on CAD/CAM systems. Figure 4.24 shows some of the sample CAD-generated tool paths and the machined sculptured part using the IDEAS^{TM} CAD/CAM system.

4.10   PROBLEMS

1. Interpret the following NC program block by block. Indicate G and M functions for each block, and plot the complete tool center path with the corresponding coordinates. Mark the corresponding NC blocks on the tool path segments.

    N010 G90 G70 M03 S1200 T05
    N020 G00 X0.375 Y0.875
    N030 Z0.1
    N040 G01 Z-0.1 F10.0 M08
    N050 X2.0 F20.0
    N060 G02 X2.375 Y0.5 I0.0 J-0.375
    N070 G01 Y0.375
    N080 X3.625
    N090 Y1.25
    N100 X3.5
    N110 G02 Y1.875 I0.0 J0.375
    N120 G01 X3.625
    N130 Y2.625
    N140 X2.375
    N150 Y2.5
    N160 G02 X2.0 Y2.125 I-0.375 J0.0
    N170 G01 X0.375
    N180 Y0.875
    N190 Z0.15
    N200 G00 Z2.0 M09 M05
    N210 M30

2. Interpret the following NC program block by block. Indicate G and M functions for each block, and plot the complete NC tool path. Mark the corresponding NC blocks on the tool path segments.

N010 G90 G71 M03 S1200 T01
N020 G00 X40.0
N030 G01 X155.0 F360 M08
N040 Y26.0
N050 G03 X125.0 Y56.0 I-30 J0
N060 G03 X110.0 Y52.0 I 0 J-30.0
N070 G01 X40.0 Y12.0
N080 Y0
N090 X0
N100 M30

3. Interpret the following NC program block by block. Indicate G and M functions for each block, and plot the complete NC tool center path with corresponding coordinates. Mark the corresponding NC blocks on the tool path segments.

N010 G90 G71 M03 S1200 T01
N020 G00 Z7
N030 G01 Z0 F100 M08
N040 X30
N050 X90 Y10
N060 Y40
N070 G02 G91 X-30 Y30 I0 J30
N080 G01 X-40
N090 G03 G90 X0 Y50 I+10 J-30
N100 G01 Y0
N110 Z7
N120 G00 X30 Y40.0
N140 G01 Z-2.5
N150 Z7
N160 G00 X0 Y0 Z20 M05 M02

4. Draw flowcharts of NC program generation processes in APT language and CAD/CAM (graphics) systems. What are the main differences between the two methods?

5. Examine the following part program and sketch the geometry of the component and cutter path.

PARTNO/ Assignment
POST/UBC-CNC-M1
CLPRINT
UNITS/MM

TOLER/0.005
P0=POINT/-20,0
P1=POINT/90,15
P2=POINT/0,0
P3=POINT/90,0
P4=POINT/30,75
L1=LINE/P2,P3
C1=CIRCLE/CENTER,P1,RADIUS,15
L2=LINE/P4,LEFT,TANTO,C1
L3=LINE/P2,P4
PL1=PLANE/P2,P3,P4
CUTTER/15
SPINDL/360
FEDRAT/50
COOLNT/ON
FROM/P0
GO/TO,L1,TO,PL1,TO,L3
GORGT/L1,TANTO,C1

Figure 4.25: The tool path coordinates are given for an end mill with 0.25″ diameter cutter.

GOFWD/C1,PAST,L2
GOFWD/L2,PAST,L3
GOLFT/L3,PAST,L1
RAPID
GOTO/P0
COOLNT/OFF
FINI

6. Write manual NC and APT codes for the part shown in Figure 4.25.

CHAPTER FIVE

DESIGN AND ANALYSIS
OF CNC SYSTEMS

5.1 INTRODUCTION

A diagram of a typical three-axis CNC machining center is shown in Figure 5.1. The CNC machining center consists of mechanical, power electronic, and CNC units. The mechanical unit consists of beds, columns, spindle assembly, and feed drive mechanisms. Spindle and feed drive motors and their servoamplifiers, high voltage power supply unit, and limit switches are part of the power electronics group. The CNC consists of a computer unit and position and velocity sensors for each drive mechanism. The operator enters the NC program to the CNC unit. The CNC computer processes the data and generates discrete numerical position commands for each feed drive and velocity command for the spindle drive. The numerical commands are converted into signal voltage ($\pm 5\,\text{V}$ or $\pm 10\,\text{V}$) by the CNC unit and sent to servoamplifiers, which process and amplify them to the high voltage levels required by the drive motors. As the drives move, sensors measure their velocity and position. The CNC periodically executes digital control laws that maintain the feed speed and tool path at programmed rates by using sensor feedback measurements.

The fundamental principles of designing CNC systems are covered in this chapter. First, sizing and selecting drive motors are presented, followed by physical structure and modeling of a servodrive control system. The mathematical modeling and analysis of drive systems are covered both in the time and frequency domain. Interpolation techniques for multiaxis contouring are presented. The chapter includes sample CNC design examples from real life.

5.2 MACHINE TOOL DRIVES

The drives in machine tools are classified as spindle and feed drive mechanisms. Spindle drives rotate over a wide velocity range (i.e., up to 35,000 rev/min), whereas the feed drives usually convert angular motions of the motors to linear traverse speeds, which can range up to 30,000 mm/min. In this text, the servocontrol of feed drives is covered only, although the material can be easily extended to spindle drives since the fundamental design and analysis methods are quite similar.

Let us take one of the feed drives in a machine tool as an example. The feed drive has the following mechanical components: the table with a workpiece,

Figure 5.1: Functional diagram of a 3-axis CNC machining center.

the nut, the ball lead screw, the torque reduction gear set, and the servomotor (see Fig. 5.2). Because of their efficient torque delivery capability at various speeds, the most common servomotors used in the feed drives are direct current (dc) motors. However, alternating current (ac) servomotors have also gained popularity owing to their improved performance. The electrical components of a servomotor system comprise the servomotor amplifier, velocity and position feedback transducers, a digital computer, and a digital to analog converter circuit.

5.2.1 Mechanical Components and Torque Requirements

The feed drive motor has to overcome both the static and dynamic loads in the machine tool. The sources of the static loads are the friction losses in the guideways and bearings and the cutting forces acting in the feeding direction of the table. The motor must deliver a high enough dynamic torque to accelerate the table, workpiece, and leadscrew assembly for a short period of time until the drive reaches the desired steady state speed. The dynamic torque is given as *peak torque* or *peak current* delivery with a period of 2–3 seconds by the servomotor manufacturers. The motors must have a sufficiently high continuous torque delivery range and a sufficient peak torque and delivery period to overcome the static and dynamic loads respectively. Estimation of static and dynamic motor loads are briefly introduced in the following paragraphs.

Static Loads. There are three sources of static loads: the friction in the slideways, frictional losses in the feed drive bearings, and cutting forces.

The friction in the guideways depends on the type of contact between the sliding table and the stationary guideway. High friction coefficients are found

Figure 5.2: Feed drive mechanism with a ball screw drive.

at plain lubricated guideways where the metal to metal table–guideway surface contact area is large. The metal contact surface area is reduced in hydrostatic and hydrodynamic guideways where pressurized lubricant is injected between the table and the guideway. The friction coefficient is probably smallest in the guideway designs where roller bearings are used at the guideway–table assembly (see Fig. 5.2). The torque reflected on the feed drive motor due to friction (T_{gf}) in the guideways can be estimated as

$$T_{gf} = \frac{h_p}{2\pi} \mu_{gf} \left[(m_t + m_w)g + F_z \right], \tag{5.1}$$

where μ_{gf} is the friction coefficient on the guideways, m_t is the table mass, m_w is the workpiece mass, F_z is the normal cutting force on the table, h_p is the leadscrew pitch length, and g is the gravitational acceleration ($9.81 \, \text{m/s}^2$).

The friction coefficient for plain guideways is typically ranges from 0.05 to 0.1 and the vertical cutting force (F_z) can be taken as 10% of the maximum resultant cutting force in a typical vertical milling machine tool [71].

Axial thrust bearings are used at both ends of the leadscrew in order to absorb the feed forces and also to guide the screw radially at the same time [72]. Axial thrust bearings are preloaded in tension to offset backlash produced by the thermal expansion of the leadscrew due to friction in the feed drive assembly. In addition to preload, the axial thrust bearings are loaded by the feed forces. The feed forces can be estimated using the cutting mechanics relationships given in the metal cutting chapter. The torque lost in the bearings and preload is estimated as

$$T_{lf} = \mu_b \frac{d_b}{2} (F_f + F_p), \tag{5.2}$$

where μ_b is the friction coefficient on bearings (typically around 0.005), d_b is the mean bearing diameter, F_f is the maximum feed force on the table, and F_p is the preload force.

The torque reflected on the leadscrew shaft due to the cutting forces in the feed direction is given as

$$T_f = \frac{h_p}{2\pi} F_f. \tag{5.3}$$

The total static disturbance load reflected on the leadscrew shaft (T_s) is found by summing the three torque values calculated in Eqs. (5.1), (5.2), and (5.3):

$$T_s = T_{gf} + T_{lf} + T_f. \tag{5.4}$$

In the cases where the static torque (T_s) is too large, a gear reduction can be applied between the motor shaft and the leadscrew. The gear reduction ratio (r_g) is defined as

$$r_g = \frac{z_l}{z_m} = \frac{n_m}{n_l}, \tag{5.5}$$

where z_m is the number of teeth on the motor's gear, z_l is the number of teeth on the feed screw's gear, n_m is the motor's angular velocity (rev/min), and n_l is the feed screw's angular velocity (rev/min).

To reduce the speed, we must have $z_l > z_m$, which gives a gear reduction ratio larger than one (i.e., $r_g > 1$). The reduced reflected torque on the motor's shaft (T_{sr}) is found as

$$T_{sr} = \frac{T_s}{r_g}. \tag{5.6}$$

The CNC designer must select a dc motor that has a larger continuous torque delivery capacity than the static torque reflected on the motor's shaft.

Dynamic Loads. Machine tools require high acceleration torque during speed changes. The reflected inertia on the motor's shaft consists of the inertia of the table, the workpiece, the leadscrew, the gears, and the motor's shaft. The moment of inertia of the table and workpiece reflected on the leadscrew shaft is

$$J_{tw} = (m_t + m_w) \left(\frac{h_p}{2\pi} \right)^2. \tag{5.7}$$

The moment of inertia of the leadscrew with a pitch diameter of d_p is

$$J_l = \frac{1}{2} m_l \left(\frac{d_p}{2} \right)^2, \tag{5.8}$$

where m_l is the mass of the lead screw shaft. The total inertia reflected on the motor's shaft is

$$J_e = \frac{J_{tw} + J_l}{r_g^2} + J_m, \tag{5.9}$$

where J_m is the inertia of the motor's shaft and $r_g \geq 1$ is the gear reduction ratio between the feedscrew and motor speeds.

There is another friction torque in the drive system that is proportional to the velocity, namely the viscous friction torque. The total dynamic torque required to accelerate the inertia J_e and to overcome viscous friction and the static loads is given as

$$T_d = J_e \frac{d\omega}{dt} + B\omega + T_{sr}, \tag{5.10}$$

where ω is the angular velocity of the motor and B is the viscous friction coefficient. Note that since the cutting is carried out at low feeds, the contribution of cutting forces to the static torque (T_s) does not have to be considered in Eq. (5.10). The peak torque value delivered by the motor must be larger than the dynamic torque calculated from Eq. (5.10). If a gear reduction is used between the motor's shaft and leadscrew, the dynamic torque reflected on the motor's shaft is reduced; see Eqs. (5.9) and (5.10).

Example. A vertical milling machine is to be retrofitted with three identical dc servomotors. Since the largest load is applied in the longitudinal axis, the motors are selected according to torque requirements of this axis. The following parameters are given for the feed drive axis:

$m_t = 180$ kg – table mass,
$m_w = 200$ kg – maximum mass for the workpiece,
$m_l = 8.15$ kg – leadscrew mass,
$h_p = 0.00508$ m/rev. – pitch of the feedscrew,
$d_p = 0.0445$ m – feedscrew diameter,
$J_m = 2.373 \times 10^{-3}$ kg m^2 – motor's shaft inertia,
$r_g = 1.$ – gear reduction ratio,
$\mu_{gf} = 0.1$ – friction coefficient in the guides,
$\mu_b = 0.005$ – friction coefficient of bearings,
$F_z = 2,000$ N – maximum vertical force,
$B \approx 0.0$ Nm/(rad/s) – viscous damping coefficient,
$F_f = 8,000$ N – maximum feeding force,
$F_p = 5,000$ N – preload force in thrust bearings,
$f = 0.170$ m/s – rapid traverse velocity,
$t_r = 0.1$ s – servo rise time.

Static Torque The static torque contributed by the friction in the guideways (Eq. 5.1) is

$$T_{gf} = 0.1 \frac{0.00508}{2\pi} [(180 + 200)9.81 + 2000] = 0.4631 \,\text{Nm}.$$

The torque lost in the bearings due to friction (Eq. 5.2) is

$$T_{\text{bf}} = 0.005 \frac{0.0445}{2} (5000 + 8000) = 1.4463 \, \text{Nm}.$$

The torque required to overcome feed forces (Eq. 5.3) is

$$T_{\text{f}} = \frac{0.00508}{2\pi} 8000 = 6.4680 \, \text{Nm}.$$

Thus the total required continuous torque from the dc motor (Eq. 5.4) is

$$T_{\text{sr}} = \frac{0.4631 + 1.4463 + 6.4680}{1.0} = 8.3874 \, \text{Nm}.$$

Dynamic Load The moment of inertia of the table and workpiece reflected on the feedscrew shaft (Eq. 5.7) is

$$J_{\text{tw}} = (180 + 200)\left(\frac{0.00508}{2\pi}\right)^2 = 2.484 \cdot 10^{-4} \, \text{kg m}^2.$$

The leadscrew's inertia (Eq. 5.8) is

$$J_l = \frac{1}{2} 8.15 \left(\frac{0.0445}{2}\right)^2 = 20.174 \cdot 10^{-4} \, \text{kg m}^2.$$

Since the motor is directly connected to the leadscrew of the machine (i.e., $r_g = 1.0$), the total inertia reflected on the motor's shaft is found (Eq. 5.9) to be

$$J_e = 2.484 \cdot 10^{-4} + 20.174 \cdot 10^{-4} + 2.373 \cdot 10^{-3} = 4.64 \cdot 10^{-3} \, \text{kg m}^2.$$

The linear acceleration of the slide is calculated by dividing the rapid feeding velocity by the servo's rise time:

$$a_l = \frac{0.170}{0.1} = 1.70 \, \text{m/s}^2.$$

The angular acceleration of the motor's shaft is

$$\frac{d\omega}{dt} = \frac{a_l}{(h_{\text{p}}/2\pi)} = \frac{1.7}{0.00508} 2\pi = 2103 \, \text{rad/s}^2.$$

The dynamic torque required is found from Eq. (5.10):

$$T_{\text{d}} = 4.64 \cdot 10^{-3} \, \text{kg m}^2 \times 2103 \, \text{rad/s}^2 + (8.3874 - 6.468) \, \text{Nm} = 11.6773 \, \text{Nm}.$$

Thus the selected servomotor must be able to attain 11.6773 Nm dynamic torque for a period of at least 0.1 s.

5.2.2 Feedback Devices

There are two basic feedback devices in the feed drive control system: position and velocity feedback transducers. Tachometers are used as the velocity transducers and encoders are used as the position feedback transducers in general [66].

Tachometers

A tachometer is a small permanent magnet dc motor mounted directly on the rear of the servomotor's shaft. The tachometer produces a voltage proportional to the actual velocity of the motor shaft. It has a factory set constant and an adjustable gain that enables tuning of the velocity feedback loop. The transfer function between the actual motor velocity and the tachometer circuit output is given as

$$\frac{V_t(s)}{\omega(s)} = T_g \cdot H_g, \tag{5.11}$$

where $V_t(s)$ is the output voltage of the tachometer circuit, $\omega(s)$ is the actual angular velocity of the motor shaft, H_g is the tachometer constant, T_g is the adjustable tachometer gain, and s is the laplace operator.

Encoders

Encoders are used as digital position measurement transducers in servo-drives. The encoder is based on the principle of emitting light with photodiodes. The encoder, which can be a disk or linear scale type, contains bands of dark and transparent segments. Light is transmitted from one side of the transparent band to the other side where there is a photodiode receiver. The photodiode gives a logic signal (i.e., a binary code) depending on the number of dark and transparent bands detected at incremental positions of the encoder. Linear encoders are used to measure the actual position of the table, whereas rotary shaft encoders are used to measure the angular position of the motor's shaft. In precision machines, a linear encoder may be used for more accurate measurement of the actual position of the table. However, if there is an unmodeled backlash in the feed drive system, the table-mounted linear encoder may produce a *limit cycle* in the position control servo. Since disk encoders are mounted directly on the motor shaft, they do not feel the backlash and therefore they do not produce a nonlinear limit cycle. Manufacturers of servomotors usually provide tachometer and shaft encoders installed on the rear of the motor's shaft in the factory. Encoder is provided with a *line density* and the type of sensing decoder used. For example, a shaft encoder with 1,000 line density and quadrature sensing decoder circuit gives 4,000 *counts* or impulses per revolution of the shaft. Encoders are simply represented as a gain K_e (counts/rad or counts/mm) in position control loop analysis.

Example. A 1,000 line encoder with quadrature sensing decoder is used as a position feedback transducer in a feed drive control system. The motor is directly connected to the leadscrew, which has a pitch length of 5.08 mm. The encoder gain is calculated as

$$K_e = \frac{4 \cdot 1000}{2\pi} \quad \text{(counts/rad)}$$

or

$$K_e = \frac{5.08}{4000} = 0.00127 \,\text{mm} \,(0.00005 \,\text{in})/\text{count}.$$

Thus, one count signal sent by the encoder corresponds to 0.00127 mm linear movement of the table. Henceforth, *count* is used to represent the basic length unit of the position control servosystem.

5.2.3 Electrical Drives

The feed drives can be powered by either electrical motors (step, dc, or ac motors) or hydraulic motors. The type of motor used depends very much on the torque delivery and time response requirements of the machine tool or robot drive.

Hydraulic motors are used when a wide torque range and rapid response are required from a drive system. Rotary hydraulic motors are used in heavy duty industrial robots, lathes, and milling machines. Linear hydraulic motors with piston-displacement units are used in grinding machines, planers, and shapers where reciprocating motion is required. The disadvantages of hydraulic drives include leakage, low efficiency, the sensitivity to dirt in the supply oil, and high maintenance costs.

Electrical stepping motors are not common in feed drive systems where the cutting load is high. They are used without feedback devices. The step motor is controlled with angular step motions that are sent in the form of control pulses from a computer. At startups and braking, if the required dynamic torque is large, the step motor may slip some of the position control pulses. Since there is no feedback device, the accuracy in machining is sacrificed. Usually, light demonstration CNC machines and material handling units employ electrical stepping motors.

The use of alternating current (ac) motors is also common in CNC machines. The speed of the ac motor is controlled by varying the frequency of the supply voltage. The fundamental problem has been to design the low cost inverters used to vary the frequency of the supply voltage. However, recent microprocessor technology has enabled the calculation of firing frequencies used in the inverter.

The most common motors used in the feed drives are direct current (dc) motors since they allow a wide range of operating speeds with the sufficiently large torque delivery required by machine tools and robots. In the following section, analysis of a permanent magnet dc motor is explained to provide a detailed understanding of the control of dc motors. However, the analysis and modeling procedures are similar for both dc and ac servomotors.

Figure 5.3: Electrical diagram of a dc motor.

5.2.4 Permanent Magnet Armature-Controlled dc Motors

An electrical diagram of a permanent magnet dc motor is shown in Figure 5.3. The speed of the dc motor is controlled by feeding a dc voltage V_a to the armature (rotor) of the motor. The dc voltage produces a variable dc current I_a on the armature, which in turn creates a magnetic field between the rotor and stationary stator. Note that the current drawn by the armature can not exceed the maximum current supply capacity of the power amplifier. The current limit is treated as a nonlinearity in the dc motor control system. It is evident that peak current will be drawn for a short period during acceleration and braking of the motor. Peak current and its duration time are given by the amplifier manufacturers. Magnetic flux is constant while the armature voltage is variable in armature-controlled dc motors. The magnetic field produces a torque, which is used for rotating the rotor connected to the motor's shaft. In dc motors, a back electromotor voltage (e.m.v.) V_b, which is proportional to the rotor speed, is produced in the rotor circuit.

The following fundamental dynamic equations govern the motion of dc motors.

The applied armature voltage V_a is derived by applying Kirchoff's law to the motor's circuit:

$$V_a(t) = R_a I_a(t) + L_a \frac{dI_a(t)}{dt} + K_b \omega(t), \tag{5.12}$$

where ω is the angular velocity (rad/s), I_a is the armature current (A), R_a is the armature resistance (Ω), L_a is the armature inductance (H), K_b is the motor's voltage (e.m.f.) constant (V/(rad/s)), and t is the time (s).

The magnetic field produces a useful motor torque T_m, which is proportional to the armature current I_a:

$$T_m(t) = K_t\, I_a(t). \tag{5.13}$$

The useful torque produced by the motor is spent in accelerating the reflected inertia on the motor's shaft, overcoming the friction in the motor's bearings and guideways, and resisting against feed cutting forces and friction loads reflected as a disturbance torque on the motor's shaft. Thus

$$T_m(t) = J_e \frac{d\omega(t)}{dt} + B\,\omega(t) + T_s(t), \tag{5.14}$$

where J_e is the reflected inertia on the motor shaft, B is the equivalent friction (viscous damping), and T_s is the static disturbance torque reflected on the motor shaft.

Note that the disturbance torque consists of cutting and friction components. The Coulomb friction torque opposes the velocity (ω). The direction of the velocity is considered in the block diagram (Fig. 5.4) using the sign function sgn [73]. In the Eq. (5.14) the viscous friction is assumed to be proportional to the speed. The behavior of the friction torque varies depending on the type of guideways used in the machine tool. The viscous friction torque is linearly proportional to the velocity, and the proportionality constant is B.

Note that Coulomb type sticking friction is usually dominant in feed drives. Coulomb friction demands constant friction torque (or current) and is independent of the feeding velocity variation. The Coulomb friction torque and viscous friction coefficient can be identified from Eq. (5.14) by measuring the armature

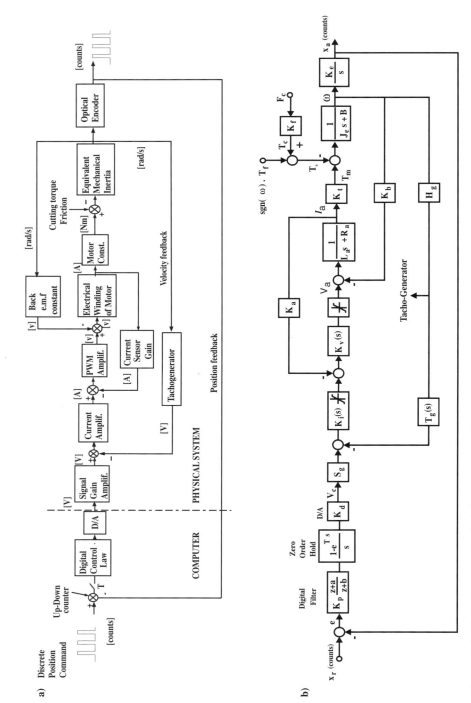

Figure 5.4: Block diagram of a feed drive servocontrol system. a) Physical parts. b) Transfer function.

current at various steady state feeding velocities (i.e., $d\omega/dt = 0$). The dc component of the resulting curve gives the constant current drawn by Coulomb friction, and the slope of the linear component is the viscous damping coefficient (B). To avoid nonlinearity in the system, Coulomb friction is disregarded and an average viscous damping value is assumed.

Taking the Laplace transforms of Eqs. (5.12), (5.13), and (5.14) gives

$$I_a(s) = \frac{V_a(s) - K_b\,\omega(s)}{L_a\,s + R_a},$$

$$T_m(s) = K_t\,I_a(s), \tag{5.15}$$

$$\omega(s) = \frac{T_m(s) - T_s(s)}{J_e\,s + B}.$$

These equations have physical interpretations. The transfer function of the armature current is derived using the error voltage as an input. The error voltage is the difference between the supplied reference armature voltage V_a and back e.m.v., which acts like a feedback signal. The resulting current produces a useful torque $T_m(s)$. A portion of this torque is spent in overcoming the disturbance torque $T_s(s)$. The remaining torque is used to accelerate the inertia and overcome the viscous friction torque, which is proportional to the velocity. The two important design parameters are

$$\tau_e = \frac{L_a}{R_a},$$

$$\tau_m = \frac{J_e}{B}, \tag{5.16}$$

where τ_m is the mechanical time constant and τ_e is the electrical time constant of the dc motor. Both time constants are provided by the motor's manufacturer by considering only the inertia of the motor's shaft and friction in the motor's shaft bearings. The mechanical time constant of the dc motor obviously increases when the reflected inertia of the table and leadscrew assembly are taken into account.

dc Motor Power Amplifiers and Velocity Control Loop

The armature voltage is supplied by a power amplifier. The power amplifier receives a large constant dc voltage from a transformer, which converts ac line voltage to the desired dc voltage level. The power amplifier illustrated in this text is a pulse width modulated (PWM), current-controlled amplifier. However, the operation of other types of power amplifiers are quite similar.

The block diagram of a complete velocity control loop is shown in Figure 5.4. The power amplifier receives a velocity command signal V_c from the digital to analog converter output of the digital controller. The velocity command signal is first buffered with an adjustable differential preamplifier represented by a gain S_g. The output of S_g is compared with an actual velocity signal measured by the tachometer feedback unit. The resulting velocity error signal (in volts) is converted to a demand current by the current amplifier, which has a gain of K_I. Most of the current amplifiers use an armature current feedback signal to improve the dynamic response of the motor. The feedback current signal is pulled from a current sense coupler and compared with the demand current. The PWM circuit generates a varying saw-tooth shape dc voltage at certain frequency. The PWM frequency is usually above 10 khz. Low frequency PWM switching signals (up to 6 khz) generate irritating audible noise. The current error signal is modulated by the PWM circuit, which is modeled as a gain K_v. The resulting dc voltage becomes an ON–OFF type rectangular waveform. The average voltage level of the waveform (dc value) is used as armature voltage V_a in the calculations. The complete block diagram of the amplifier, motor, and tachometer feedback unit are shown in Figure 5.4a.

The block diagram is organized with temporary state variables to illustrate the derivation of the velocity loop transfer function. The following relationships can be expressed from the block diagram using temporary states, V_1, V_2, and V_3:

$$V_1(s) = S_g V_c(s) - T_g H_g \omega(s),$$
$$\begin{aligned} V_2(s) &= K_{rmI} V_1(s) - K_a I_a(s) \\ &= K_I S_g V_c(s) - K_I T_g H_g \omega(s) - K_a I_a(s), \end{aligned}$$
$$\begin{aligned} V_3(s) &= K_v V_2(s) - K_b \omega(s) \\ &= K_v K_I S_g V_c(s) - (K_v K_I T_g H_g + K_b)\omega(s) - K_v K_a I_a(s). \end{aligned}$$

The transfer function between the current and state V_3 is

$$I_a(s) = \frac{V_3(s)}{L_a s + R_a}.$$

Substituting the value of V_3 into the current expression yields

$$I_a(s) = \frac{K_v K_I S_g}{L_a s + R_a + K_v K_a} V_c(s) - \frac{K_v K_I T_g H_g + K_b}{L_a s + R_a + K_v K_a} \omega(s). \tag{5.17}$$

The motor's mechanical transfer function (see Eq. 5.15) is

$$\omega(s) = \frac{T_m(s) - T_s(s)}{J_e s + B}$$

or

$$\omega(s) = \frac{K_t}{J_e s + B} I_a(s) - \frac{1}{J_e s + B} T_s(s). \tag{5.18}$$

Substituting the current expression (5.17) into Eq. (5.18) yields the transfer function between the output velocity ω and the velocity command input voltage V_c and the disturbance torque T_d:

$$\omega(s) = \frac{K_1}{s^2 + K_2 s + K_3} V_c(s) - \frac{(1/J_e)[s + (R_a + K_v K_a)/L_a]}{s^2 + K_2 s + K_3} T_s(s), \tag{5.19}$$

where

$$K_1 = \frac{K_t S_g K_I K_v}{L_a J_e},$$

$$K_2 = \frac{B}{J_E} + \frac{R_a + K_v K_a}{L_a},$$

$$K_3 = \frac{B(R_a + K_v K_a) + K_t(K_b + H_g T_g K_v K_I)}{J_e L_a}$$

The feed drive servo velocity controller is designed to have a fast rise time with zero overshoot at step changes in the velocity. Let us analyze the velocity loop as a function of the velocity command input voltage V_c. The transfer function (Eq. 5.19) can be expressed as

$$\frac{\omega(s)}{V_c(s)} = \frac{K_1}{s^2 + 2\,\xi\,\omega_n\,s + \omega_n^2}. \qquad (5.20)$$

Here the natural frequency (ω_n) and damping ratio (ξ) of the velocity loop are defined as follows:

$$\omega_n = \sqrt{K_3} \quad [\text{rad/s}],$$
$$\xi = \frac{K_2}{2\sqrt{K_3}} < 1, \qquad (5.21)$$

where $K_1, K_2 > 0$. The time domain step response of this underdamped velocity servo is expressed as

$$\omega(t) = V_c \cdot \frac{K_1}{K_3}\left[1 - \frac{e^{-\xi\omega_n t}}{\sqrt{1-\xi^2}} \cdot \sin(\omega_d\,t + \phi)\right], \qquad (5.22)$$

where damped natural frequency ω_d and phase shift ϕ are defined as

$$\omega_d = \omega_n\sqrt{1-\xi^2},$$
$$\phi = \tan^{-1}\left(\frac{\sqrt{1-\xi^2}}{\xi}\right).$$

The variable gains of the amplifier (i.e., S_g, T_g, K_I, K_v) are tuned to have a desired velocity loop gain and step response characteristics as shown in Figure 5.5. When a unit step input (i.e., $V_c = 1\,\text{V}$) is applied on the amplifier input port, the maximum response of the velocity loop occurs at time t_p where the derivative of the velocity is zero (i.e., $d\omega_n(t)/dt = 0$). At the first overshoot, the following expression can be obtained from the time derivative of Eq. (5.22):

$$t_p = \frac{\pi}{\omega_d}. \qquad (5.23)$$

Typical design values for a feed drive servo may be a damping ratio of $\xi = 0.707$ and a peak time of $t_p = 10\,\text{ms}$. The natural frequency ω_n can be estimated from Eq. (5.23):

$$\omega_n = \frac{\pi}{t_p\sqrt{1-\xi^2}}$$
$$= 554\,\text{rad/s} = 88\,\text{Hz}.$$

Figure 5.5: Step response of a second-order underdamped system (τ = time constant, t_p = peak time, t_r = rise time, M_p = overshoot).

The corresponding servo parameters are identified by substituting the values of ξ and ω_n into Eq. (5.21). It is fairly obvious that one can not demand a higher natural frequency than the maximum capacities of the motor and amplifier gains can provide.

5.2.5 Position Control Loop

The position loop consists of an up–down counter, an encoder, a digital compensation filter, and a digital to analog converter.

Up–Down Counter

The up–down counter register receives both command and measured position counts. The position commands increment while the encoder feedback counts decrement the contents of the up–down counter. The instantaneous content of the counter represents the accumulated or integrated position error within digital servo control interval T. The block diagram representation of the up–down counter is

$$\frac{X_a(s)}{\omega(s)} = \frac{K_e}{s}.$$
(5.24)

Digital Compensation Filter

The contents of the up–down counter, the position error, is sampled every T seconds. The discrete position error $E(k)$ is the difference between the reference and actual position of the table:

$$E(k) = X_r(k) - X_a(k),$$
(5.25)

where $X_r(k)$ is the discrete reference position, $X_a(k)$ is the discrete actual position, and z is the discrete forward time shift operator.

The error is passed from a digital filter, which has a typical transfer function form of

$$D(z) = K_p \frac{z + a}{z + b},$$
(5.26)

where K_p is the position control filter gain, a is the filter's zero, and b is the filter's pole.

The digital filter is programmable and resides in the servo motion control computer. The filter's parameters are tuned to provide a desired transient response of the position control system.

Digital to Analog (D/A) Converter

The motion control computer sends the digital filter's output to a D/A converter circuit of the motion control computer. The D/A converter is modeled as a zero order hold (ZOH) and a gain K_d. The D/A gain is found as

$$K_d = \frac{\text{voltage range of the D/A chip}}{2^{nb}},$$
(5.27)

where nb is the number of bits employed by the D/A chip in converting a binary number to analog voltage. For example; a 12 bit D/A converter chip with $\pm 10\,\text{V}$ voltage range has a gain of

$$K_{\text{d}} = \frac{20\,\text{V}}{2^{12}} = 0.00488 \text{ (V/count)}.$$

5.3 TRANSFER FUNCTION OF THE POSITION LOOP

The block diagram of the complete position control system can be organized as shown in Figure 5.4. The system has continuous and discrete components. The continuous part of the system's transfer function is represented in Laplace domain as

$$G_{\text{c}}(s) = \frac{K_1}{s^2 + K_2\,s + K_3}\,\frac{K_{\text{e}}}{s}. \tag{5.28}$$

The velocity control signal V_{c} of the digital motion control unit is applied to the power amplifier at T second intervals via a D/A converter with a gain of K_{d}. The zero-order hold equivalent of the $G_{\text{c}}(s)$ for one millisecond sampling interval is [74]

$$G_{\text{c}}(z) = K_{\text{d}}(1 - z^{-1})\,\mathcal{Z}\left[\frac{G_{\text{c}}(s)}{s}\right],$$

which, after taking the z transform, becomes

$$G_{\text{c}}(z) = \frac{K_{\text{d}}K_1 K_{\text{e}}}{K_3}\,\frac{b_2 z^2 + b_1 z + b_0}{(z-1)(z^2 + a_1 z + a_0)}, \tag{5.29}$$

where the forward operator is equivalent to $z = e^{sT}$. The derived expressions for the parameters of $G_{\text{c}}(z)$ are given as

$$b_2 = T - \frac{1}{\omega_{\text{d}}}\,e^{-\xi\omega_{\text{n}}T}\,\sin(\omega_{\text{d}}T)$$

$$- \frac{K_2}{K_3}\left\{1 - e^{-\xi\omega_{\text{n}}T}\left[\frac{\xi\omega_{\text{n}}}{\omega d}\,\sin(\omega_{\text{d}}T) + \cos(\omega_{\text{d}})\right]\right\},$$

$$b_1 = 2\,e^{-\xi\omega_{\text{n}}T}\left[\frac{\sin(\omega_{\text{d}}T)}{\omega_{\text{d}}} - T\cos(\omega_{\text{d}}T)\right]$$

$$+ \frac{K_2}{K_3}(1 - e^{-2\xi\omega_{\text{n}}T}) - 2\,e^{-\xi\omega_{\text{n}}T}\,\sin(\omega_{\text{d}}T)\,\frac{K_2}{K_3}\,\frac{\xi\omega_{\text{n}}}{\omega_{\text{d}}},$$

$$b_0 = Te^{-2\xi\omega_{\text{n}}T} - \frac{1}{\omega_{\text{d}}}\,e^{-\xi\omega_{\text{n}}T}\,\sin(\omega_{\text{d}}T)$$

$$+ \frac{K_2}{K_3}\left\{e^{-2\xi\omega_{\text{n}}T} + e^{-\xi\omega_{\text{n}}T}\left[\frac{\xi\omega_{\text{n}}}{\omega_{\text{d}}}\,\sin(\omega_{\text{d}}T) - \cos(\omega_{\text{d}}T)\right]\right\},$$

$$a_1 = -2\,e^{-\xi\omega_{\text{n}}T}\,\cos(\omega_{\text{d}}T),$$

$$a_0 = e^{-2\xi\omega_{\text{n}}T}.$$

Figure 5.6: Discrete time position control loop.

The reduced block diagram of the equivalent discrete position control system is shown in Figure 5.6. The closed-loop transfer function of the complete feed drive control system is derived as follows:

$$G_{cl}(z) = \frac{X_a(k)}{X_k(z)} = \frac{D(z)\,G_c(z)}{1 + D(z)G_c(z)},$$

or

$$G_{cl}(z) = K_{cl}\,\frac{z^3 + \beta_2 z^2 + \beta_1 z + \beta_0}{z^4 + \alpha_3 z^3 + \alpha_2 z^2 + \alpha_1 z + \alpha_0}. \tag{5.30}$$

The parameters of the closed-loop transfer function $G_{cl}(z)$ are given as follows:

$$\beta_2 = \frac{b_1 + a\,b_2}{b_2},$$

$$\beta_1 = \frac{b_0 + a\,b_1}{b_2},$$

$$\beta_0 = \frac{a\,b_0}{b_2},$$

$$\alpha_3 = \frac{K_1\,K_e\,K_d\,K_p}{K_3}\,b_2 + b + a_1 - 1,$$

$$\alpha_2 = \frac{K_1\,K_e\,K_d\,K_p}{K_3}\,(b_1 + a\,b_2) + b(a_1 - 1) + a_0 - a_1,$$

$$\alpha_1 = \frac{K_1\,K_e\,K_d\,K_p}{K_3}(b_0 + a\,b_1) + b(a_0 - a_1) - a_0,$$

$$\alpha_0 = \frac{K_1\,K_e\,K_d\,K_p}{K_3}\cdot a\cdot b_0 - ba_0,$$

$$K_{cl} = \frac{K_1\,K_e\,K_d\,K_p\,b_2}{K_3}.$$

The transfer function (Eq. 5.30) can be expressed using the backward time shift operator z^{-1} by multiplying both numerator and denominator by z^{-4}. This gives

$$G_{cl}(z^{-1}) = K_{cl}\,\frac{z^{-1} + \beta_2 z^{-2} + \beta_1 z^{-3} + \beta_0 z^{-4}}{1 + \alpha_3 z^{-1} + \alpha_2 z^{-2} + \alpha_1 z^{-3} + \alpha_0 z^{-4}} \tag{5.31}$$

Note that the z^{-1} backward time shift operator acts on the discrete signal as

$$z^{-1}x(kT) = x[(k-1)T],$$

where T is the discrete sampling time interval and $x(kT)$ is the discrete value of x at sampling interval number k.

For any given discrete time input $X_r(kT)$, the position response $X_a(kT)$ can be found by rearranging the transfer function $G_{cl}(z^{-1})$:

$$X_a(k) = -(\alpha_3 z^{-1} + \alpha_2 z^{-2} + \alpha_1 z^{-3} + \alpha_0 z^{-4})X_a(z^{-1})$$
$$+ K_{cl} \cdot (z^{-1} + \beta_2 z^{-2} + \beta_1 z^{-3} + \beta_0 z^{-4})X_r(z^{-1}). \tag{5.32}$$

The discrete time response of the position servo can be simulated by taking the inverse z^{-1} transform of Eq. (5.32). The resulting difference equation contains the discrete history of input and the past history of output position values.

Note that the position control loop can be analyzed in the continuous time (s) domain as an alternative to the discrete time (z) domain analysis. However, this approach requires approximating the digital filter $(D(z))$ in the s domain. One of the widely used approximations is $z = (1 + sT/2)/(1 - sT/2)$, which is called Tustin's bilinear transformation. The corresponding digital filter becomes

$$D(s) = K_p' \frac{s + a'}{s + b'}, \tag{5.33}$$

where

$$K_p' = K_p \frac{1 - a}{1 - b}, \quad a' = \frac{2}{T}\frac{1 + a}{1 - a}, \quad b' = \frac{2}{T}\frac{1 + b}{1 - b}$$

or

$$K_p = K_p' \frac{1 - b'}{1 - a'}, \quad a = \frac{Ta' - 2}{Ta' + 2}, \quad b = \frac{Tb' - 2}{Tb' + 2}.$$

In this case, the closed-loop transfer function of the position loop in Laplace domain becomes

$$G_{cl}(s) = \frac{D(s)K_d G_c(s)}{1 + D(s)K_d G_c(s)}. \tag{5.34}$$

Following Error in CNC Systems

There are two critical performance requirements from the feed drive servo. The first is to obtain a smooth transient response to avoid an oscillatory tool path during velocity changes; the second is to minimize the steady state position error, the *following error*, to achieve precision multiaxis contouring. At steady state contouring with a feeding velocity f_c (counts/s), the reference position is expressed as a ramp type input

$$X_r(kT) = f_c \cdot kT, \tag{5.35}$$

where k is the sampling counter. In the z domain, the ramp command is expressed as

$$X_r(k) = f_c \frac{Tz}{(z - 1)^2}. \tag{5.36}$$

The corresponding following error is then derived as

$$e_{ss} = \lim_{z \to 1} \frac{f_c T z}{(z-1)D(z)G_c(z)}. \tag{5.37}$$

Substituting Eqs. (5.26), (5.29), and (5.35) into Eq. (5.37) gives the parametric expression for the following error in the feed drives as

$$e_{ss} = \frac{f_c K_3 (1+b)}{K_1 K_e K_p K_d (1+a)}. \tag{5.38}$$

It is evident that the higher the open-loop transfer function gain is (i.e., $D(z)G_c(z)$), the smaller the following error will be, which is desired for accurate multiaxis contouring operations. However, high open-loop gain is limited by the inertia of the mechanical drive system and the limits of motor torque and amplifier. The control engineer must tune the digital control (i.e., filter) parameters to achieve an optimum feed drive servo response without causing any oscillation or overshoot.

The steady state error can also be estimated using the continuous time domain transfer functions as

$$e_{ss} = \lim_{s \to 0} s \frac{f_c}{s^2 [D(s)K_d G_c(s)]}, \tag{5.39}$$

where $G_c(s)$ and $D(s)$ are given in Eqs. (5.28) and (5.33).

5.4 STATE SPACE MODEL OF FEED DRIVE CONTROL SYSTEMS

A state space model for the feed drive servo is used to verify the derived model using experimentally measured time domain response data. The servo is again divided into continuous and discrete sections for state space modeling [75].

The continuous part of the system consists of the velocity control loop (Eq. 5.28) and the up–down counter (Eq. 5.24). Three *states* – the armature current I_a, the angular velocity ω, and the actual position X_a – are derived as follows: From Figure 5.4, the armature voltage can be expressed as

$$V_a = K_v [K_I (S_g V_c - T_g H_g \omega) - K_a I_a]. \tag{5.40}$$

Substituting Eq. (5.12) into (5.40) yields

$$\frac{dI_a}{dt} = -\frac{K_v K_a + R_a}{L_a} I_a - \frac{K_b + K_v K_I T_g H_g}{L_a} \omega + \frac{K_v K_I S_g}{L_a} V_c. \tag{5.41}$$

Eliminating the motor torque T_m by substituting Eq. (5.13) into Eq. (5.14) gives

$$\frac{d\omega}{dt} = \frac{K_t}{J_e} I_a - \frac{B}{J_e} \omega - \frac{1}{J_e} T_s. \tag{5.42}$$

The inverse Laplace transform of the transfer function of the up–down counter and the encoder (Eq. 5.24) gives

$$\frac{dX_a(t)}{dt} = K_e \cdot \omega(t). \tag{5.43}$$

The state equations (5.41–5.43) can be organized in a standard state space form:

$$\dot{x}_c(t) = A_c\, x_c(t) + B_c\, u_c(t), \tag{5.44}$$

where the *state vector* $(x_c(t))$ and the *input vector* $(u_c(t))$ are defined as

$$x_c(t) = \begin{bmatrix} I_a(t) \\ \omega(t) \\ X_a(t) \end{bmatrix}, \quad u_c(t) = \begin{bmatrix} V_c(t) \\ T_s(t) \end{bmatrix}$$

and A_c and B_c are constant matrices:

$$A_c = \begin{bmatrix} -\dfrac{K_v K_a + R_a}{L_a} & -\dfrac{K_b + K_v K_I T_g H_g}{L_a} & 0 \\ \dfrac{K_t}{J_e} & -\dfrac{B}{J_e} & 0 \\ 0 & K_e & 0 \end{bmatrix},$$

$$B_c = \begin{bmatrix} \dfrac{K_v K_I S_g}{L_a} & 0 \\ 0 & -\dfrac{1}{J_e} \\ 0 & 0 \end{bmatrix}.$$

The state equation (5.44) represents the continuous part of the feed drive servo system, which has the following discrete equivalent solution for an observation interval of T [75]:

$$x_c(k + 1) = \Phi(T)\, x_c(k) + H(T)\, u_c(k), \tag{5.45}$$

where the state and input vectors at sampling interval k are defined as

$$x_c(k) = \begin{bmatrix} I_a(k) \\ \omega(k) \\ X_a(k) \end{bmatrix}, \quad u_c(k) = \begin{bmatrix} V_c(k) \\ T_s(k) \end{bmatrix},$$

$$\Phi(T) = e^{A_c T} = \begin{bmatrix} \phi_{11} & \phi_{12} & \phi_{13} \\ \phi_{21} & \phi_{22} & \phi_{23} \\ \phi_{31} & \phi_{32} & \phi_{33} \end{bmatrix},$$

$$H(T) = \int_0^T e^{A_c t} dt \cdot B_c = \begin{bmatrix} h_{11} & h_{12} \\ h_{21} & h_{22} \\ h_{31} & h_{32} \end{bmatrix}.$$

The matrix $\Phi(T)$ is computed from the eigenvalues of the A_c matrix or its Taylor series expansion for the discrete-time equivalent of the continuous-time system. Because the sampling interval (T) is small, the first three terms of the following Taylor series approximation may be sufficient for most applications:

$$\Phi(T) = e^{A_c T} = [I] + [A]T + [A]^2 \frac{T^2}{2!} + \cdots. \tag{5.46}$$

The discrete-time components of the position control loop consists of digital filter $D(z)$ and D/A converter gain K_d. The velocity command signal can be expressed in the z domain as

$$V_c(k) = K_p \frac{z+a}{z+b} \cdot K_d \cdot [X_r(k) - X_a(k)]. \tag{5.47}$$

The equation can be rearranged as

$$V_c(k) = K_p \, K_d \, [X_r(k) - X_a(k)] + V_d(k), \tag{5.48}$$

where

$$V_d(k) = K_p \, K_d \, \frac{a-b}{z+b} \, [X_r(k) - X_a(k)]. \tag{5.49}$$

The new variables V_c and V_d can be treated as a fourth *state* respectively. After rearranging Eqs. (5.49) and (5.47) and taking their inverse z transforms, we obtain the following discrete-time state equations:

$$\begin{aligned} V_d(k+1) &= -b \, V_d(k) + K_p \, K_d \, (a-b) \, [X_r(k) - X_a(k)], \\ V_c(k) &= K_p \, K_d \, [X_r(k) - X_a(k)] + V_d(k). \end{aligned} \tag{5.50}$$

The discrete-time state equation (5.50) can be combined with the state equation (5.45), which represents the discrete-time equivalent of the continuous part of the feed drive servo. An algebraic rearrangement yields the following complete state equations for the feed drive servo:

$$\begin{aligned} x(k+1) &= G(T) \cdot x(k) + \Gamma(T) \cdot u(k), \\ y(k) &= C_s \cdot x(k) + D_s \cdot u(k), \end{aligned} \tag{5.51}$$

where the state, input, and output vectors are defined respectively as

$$x(k) = \begin{bmatrix} V_d(k) \\ I_a(k) \\ \omega(k) \\ X_a(k) \end{bmatrix}, \quad u(k) = \begin{bmatrix} X_r(k) \\ T_s(k) \end{bmatrix}, \quad y(k) = \begin{bmatrix} V_c(k) \\ I_a(k) \\ \omega(k) \\ X_a(k) \end{bmatrix}.$$

The state matrix $G(T)$, input matrix $\Gamma(T)$, output matrix $C_s(T)$, and transmission matrix $D_s(T)$ are defined in order as follows:

$$G(T) = \begin{bmatrix} -b & 0 & 0 & -K_p K_d (a-b) \\ h_{11} & \phi_{11} & \phi_{12} & \phi_{13} - h_{11} K_p K_d \\ h_{21} & \phi_{21} & \phi_{22} & \phi_{23} - h_{21} K_p K_d \\ h_{31} & \phi_{31} & \phi_{32} & \phi_{33} - h_{31} K_p K_d \end{bmatrix},$$

$$\Gamma(T) = \begin{bmatrix} K_p K_d (a-b) & 0 \\ h_{11} K_p K_d & h_{12} \\ h_{21} K_p K_d & h_{22} \\ h_{31} K_p K_d & h_{32} \end{bmatrix},$$

$$C_s = \begin{bmatrix} 1 & 0 & 0 & -K_p K_d \\ 0 & 1 & 0 & 0 \\ 0 & 0 & 1 & 0 \\ 0 & 0 & 0 & 1 \end{bmatrix}, \quad D_s = \begin{bmatrix} K_p K_d & 0 \\ 0 & 0 \\ 0 & 0 \\ 0 & 0 \end{bmatrix}.$$

The output vector $y(k)$ gives access to three useful dynamic parameters in the feed drive servo, namely the armature current, the angular velocity, and the position of the table for a given position command and applied cutting torque. Other states in the control system can be easily found by multiplying them with the appropriate gains according to the block diagram shown in Figure 5.4.

5.5 EXAMPLE: FEED DRIVE CONTROL SYSTEM DESIGN FOR A VERTICAL MILLING MACHINE

A three-axis vertical research milling machine was retrofitted at the author's laboratory at the University of British Columbia. The retrofitted machine tool has a 5 kW ac motor connected to a spindle gear box. The three feeding axes (x, y, and z) of the machine have recirculating ball screw drives with 60, 40, and 12 cm travel limits. All three axes are driven by permanent magnet dc servomotors powered by pulse width modulated (PWM) amplifiers. The motors, which are directly connected to the leadscrew shafts, have an 11 ms mechanical (without table) time constant and a 5 ms electrical time constant, and the amplifiers have 30 A peak and 15 A continuous current delivery capacity. The amplifiers are the same type as explained in Section 5.2.3.

A personal computer (PC) is used as a master CNC. A 32-bit Digital Signal Processing Board (DSP) with a three-axis motion control module is used to control positions of the three linear axes of the machine. Both cards reside in the AT/ISA bus. The motion control card has 24-bit programmable digital input/output (I/O) lines for logic control functions. Coolant, spindle, travel limit, emergency relay, etc. logic control signals are wired to the I/O lines to control auxiliary functions of the machine tool. The real time linear, circular, and spline interpolation algorithms, and position control of all axes, are executed in the DSP board. The discrete position commands are sent to the I/O board, which converts them into analog signal voltage via 16-bit dedicated D/A converters and sends them to the servoamplifiers. The I/O board and the real time interpolation and control codes have been developed in the author's laboratory. Additional sensor-based intelligent machining process monitoring and control modules can be added to the CNC, which was designed with an open software and hardware architecture as explained in Chapter 6. The machine tool can be used as a standard CNC machine by loading an in-house developed ISO NC language emulator into the PC. Feed, acceleration, deceleration, and digital filter parameters can be changed in real time by sending the desired values from the PC to the motion control unit [76]. This feature is particularly important and essential for adaptive control and machine tool monitoring tasks.

The block diagram of the control system is the same as shown in Figure 5.4. The constants of the dc motor, gains, and digital filter parameters are given in Table 5.1. The parameters of the velocity and position loop transfer functions [77] are calculated from Eqs. (5.19), (5.29), and (5.30) and are given in Table 5.2.

TABLE 5.1. Parameters of Feed Drive System

B	0.09 Nm/(rad/s)
H_g	0.08872 V/(rad/s)
J_e	0.0036 kg m^2
K_a	0.0643 A/A
K_b	0.3 V/(rad/s)
K_d	0.00488 V/count
K_e	636.62 counts/(rad/s)
K_I	25.5 A/V
K_p	9.2038
K_v	21.934 V/V
K_t	0.3 Nm/A
L_a	2 mH
R_a	0.4 Ω
S_g	0.0648 V/V
T_g	0.13183 V/V
a	-0.7161
b	-0.6681
p	5.08 mm
Continuous current supply = 15 A	
Peak current supply = 30 A	

The calculated damping ratio and the natural frequencies of the velocity loop are found to be

$$\omega_n = \sqrt{K_3} = 554 \ (\text{rad/s}) = 88\,\text{Hz},$$

$$\xi = \frac{K_2}{2\sqrt{K_3}} = 0.8384.$$

The step response of the velocity loop is obtained by applying $V_c = 1.0$ V to the terminals of the amplifier and simultaneously measuring the velocity from the tachogenerator. The simulated step response (Eq. 5.22) of the velocity control servo is shown in Figure 5.7. The system seems to reach to the command velocity at about 10 ms without any overshoot. The frequency response of the velocity loop is analyzed to determine the bandwidth of the controller. The magnitude ratio ($M(\omega)$) and the phase angle ($\phi(\omega)$) of the closed velocity loop can be derived from Eq. (5.19):

$$M(\omega) = 20 \ \log(K_1/K_3) - 20 \log\left[\left(1 - \omega^2/\omega_n^2\right)^2 + (2\xi\omega/\omega_n)^2\right]^{\frac{1}{2}},$$

$$\phi(\omega) = -\tan^{-1}\left[\left(2\xi\frac{\omega}{\omega_n}\right)\Big/\left(1 - \left(\frac{\omega}{\omega_n}\right)^2\right)\right]. \tag{5.52}$$

The frequency response of the velocity loop can be measured with a two-channel Fourier analyzer. The analyzer's random signal generator output is fed to the input terminal of the amplifier, and the velocity is measured from the tachogenerator output. Alternatively, harmonic (i.e., sine wave) signals at different frequencies can be given as an input, and the corresponding time delay (i.e., phase delay) and magnitude of the output can be measured. The simulated frequency response results are shown in Figure 5.7. The results indicate that the servo seems to be able to follow feeding velocity changes up to 40 Hz (250 rad/s), which is the bandwidth of the velocity loop.

TABLE 5.2. Calculated Parameters of the Feed Drive Transfer Function

$K_1 = 1.5102 \cdot 10^6$	$K_2 = 930.178$	$K_3 = 3.077 \cdot 10^5$	
$b_2 = 0.40691 \cdot 10^{-4}$	$b_1 = 1.2906 \cdot 10^{-4}$	$b_0 = 0.25539 \cdot 10^{-4}$	
$a_1 = -1.1992$	$a_0 = 0.3945$		
$\beta_2 = 2.4556$	$\beta_1 = -1.6438$	$\beta_0 = -0.4495$	
$\alpha_3 = -2.8616$	$\alpha_2 = 3.077$	$\alpha_1 = -1.4686$	$\alpha_0 = 0.261$
$K_{cl} = 0.0057$			

Figure 5.7: Step and frequency response of the feed drive's velocity loop.

In addition to the velocity loop, the position loop consists of an encoder, digital filter, and integrator (i.e., up–down counter). These are mostly digital and their transfer functions are known (see Fig. 5.4). The digital filter is tuned in such a way that the overall closed-loop position control system behaves like a second-order system with a rise time of $t_r = 30$ ms and an overshoot of $M_p = 0.1\%$. The following digital filter was identified:

$$D(z) = 9.2038 \, \frac{z - 0.7161}{z - 0.6681}.$$

The calculated values of the state space parameters of the system are

$$\Phi(T) = e^{A_c T} = \begin{bmatrix} 0.3274 & -2.1160 & 0 \\ 0.0515 & 0.8718 & 0 \\ 0.0194 & 0.6049 & 1 \end{bmatrix},$$

$$H(T) = \int_0^T e^{A_c t} dt \cdot B_c = \begin{bmatrix} 11.3754 & 0.3481 \\ 0.5532 & -0.2639 \\ 0.1271 & -0.0859 \end{bmatrix},$$

$$G(T) = \begin{bmatrix} 0.6681 & 0 & 0 & 0.0022 \\ 11.3754 & 0.3274 & -2.1160 & -0.5112 \\ 0.5532 & 0.0515 & 0.8718 & -0.0249 \\ 0.1271 & 0.0194 & 0.6049 & 0.9943 \end{bmatrix},$$

$$\Gamma(T) = \begin{bmatrix} -0.0022 & 0 \\ 0.5112 & 0.3481 \\ 0.0249 & -0.2639 \\ 0.0057 & -0.0859 \end{bmatrix},$$

$$C = \begin{bmatrix} 1 & 0 & 0 & -0.0449 \\ 0 & 1 & 0 & 0 \\ 0 & 0 & 1 & 0 \\ 0 & 0 & 0 & 1 \end{bmatrix}, \quad D = \begin{bmatrix} 0.0449 & 0 \\ 0 & 0 \\ 0 & 0 \\ 0 & 0 \end{bmatrix}.$$

The steady state position error can be calculated by substituting the servo parameter values in Eq. (5.37); thus

$$e_{ss} = 0.0083 \ f_c \ \text{[counts]}.$$

The step, ramp, and frequency response of the digital position control loop are shown in Figure 5.8.

Measurements across a range of velocities showed that the steady state error is not quite linear; this is mainly due to effect of the static friction disturbance and unmodeled part of the dynamics. An average coefficient of friction is used for a general analysis and tuning of the positional control systems. However, for precision contouring, the influence of the static friction must be compensated by a more advanced control law than the simple digital filter presented in this text.

Figure 5.8: Step, ramp, and frequency response of the feed drive's position loop.

5.6 VELOCITY COMMAND GENERATION IN CNC SYSTEMS

The CNC motion control computer receives NC blocks that contain the mode of interpolation (i.e., linear – G01, circular – G02 or G03, or spline G05), the coordinates of end point ($P_e(x_e, y_e)$), arc radius ($R = \sqrt{I_c^2 + J_c^2}$), and vector feed (f). The coordinates and derivatives of a series of points along a spline trajectory are provided for quintic spline interpolation.

The acceleration (A), deceleration (D), and jerk (J) values are either set to default values within the CNC or given by the NC programmer within the NC part program. The units are converted into *counts* in the CNC system. The tool path segment (i.e., line, arc, or spline in an NC block) is broken into N number of small segments in axis directions at interpolation time intervals of T_i.

The minimum interpolation time has to be equal to or an integer multiple of the axis position control loop closure time (T). The feed f is provided by the NC part program, and the minimum interpolation period T_{min} is set within the CNC control software. The interpolation step size is calculated as

$$\Delta u = fT_{min} \tag{5.53}$$

for set values of f and T_{min}. The step size Δu is kept constant until T_{min} or f is changed. When the feed (f) is changed during machining by a feed-override switch or a sensor-based machining process control module, Δu is kept constant but the interpolation time T_i is updated as

$$T_i = \frac{\Delta u}{f}. \tag{5.54}$$

The selection of the minimum interpolation time T_{min} is limited by the computation load and speed of the CNC motion control computer where the algorithm is executed. By using a varying interpolation period, many axes are synchronized from a one-dimensional calculation. The addition of other axes has no effect on the velocity profiling; therefore the position and velocity interpolation are decoupled. Assuming that the total displacement along an arbitrary path is L, the interpolator task is executed N times at interpolation time intervals of T_i:

$$N = \frac{L}{\Delta u}. \tag{5.55}$$

N is always rounded to the next higher even integer for computational efficiency. The feed is reduced accordingly for correction. The total number of iterations (N) is divided into a number of stages depending on the type of velocity profile used for trajectory generation. A simple to implement trapezoidal velocity profile and a complex but accurate parabolic velocity profile for high speed drives are presented here.

a) Velocity profile with zero initial feed.

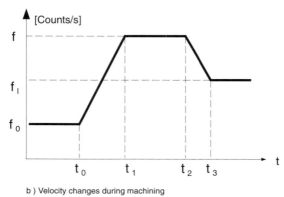

b) Velocity changes during machining

Figure 5.9: Trapezoidal velocity profile of feed.

5.6.1 Trapezoidal Velocity Profile

The acceleration and deceleration of the axis is controlled by imposing a trapezoidal velocity profile shown in Figure 5.9 on the position command generation algorithm [78]. The trapezoidal velocity profile is simple to implement, computationally advantageous, and suitable for most low speed, low cost machines. The total number of interpolation steps (N) is divided into an acceleration (N_1), constant velocity (N_2), and deceleration (N_3) zones according to the trapezoidal velocity profile shown in Figure 5.9, that is, $N = N_1 + N_2 + N_3$. The counters for acceleration and deceleration distances (N_1, N_3) are calculated from the target feed f [count/s], acceleration A [count/s^2], deceleration D [count/s^2], and displacement step Δu using the trapezoidal velocity profile. If the initial feed is zero, the total tool path length (l_1) traveled during the acceleration period ($0 < t < t_1$) can be found from Figure 5.9a as,

$$l_1 = \int_0^{t_1} At\,dt = \frac{At_1^2}{2}. \tag{5.56}$$

Since $t_1 = f/A$ for a constant acceleration, the number of interpolation intervals during acceleration is

$$N_1 = \frac{l_1}{\Delta u} = \frac{f^2}{2A\Delta u}. \tag{5.57}$$

Similarly, the deceleration counter is found as

$$N_3 = \frac{l_3}{\Delta u} = \frac{f^2}{2D\Delta u}. \tag{5.58}$$

In some NC machining applications, it is not desired to bring the machine to a full stop before switching to the next tool path segment. Alternatively, the operator or sensor-based adaptive process control algorithms may manipulate the feed within an NC block. Let the CNC system accelerate from feed f_0 to the new feed command f. The acceleration counter can be found as (Figure 5.9b)

$$l_1 = \int_{t_0}^{t_1} [f_0 + A(t - t_0)]dt = \int_0^{\tau_a} [f_0 + A\tau]d\tau = f_0\tau_a + \frac{A\tau_a^2}{2}, \tag{5.59}$$

where $\tau_a = t_1 - t_0 = (f - f_0)/A$ and

$$l_1 = \frac{f^2 - f_0^2}{2A},$$ (5.60)

which leads to

$$N_1 = \frac{l_1}{\Delta u} = \frac{f^2 - f_0^2}{2A \cdot \Delta u}.$$ (5.61)

Similarly, if the system decelerates from feed f to f_l, then

$$l_3 = \int_{t_2}^{t_3} [f - D(t - t_2)]dt = f_0\tau_d - \frac{D\tau_d^2}{2} = \frac{f^2 - f_l^2}{2D},$$ (5.62)

which leads to the following number of interpolation periods during deceleration:

$$N_3 = \frac{l_3}{\Delta u} = \frac{f^2 - f_l^2}{2D \cdot \Delta u}.$$ (5.63)

The counters N, N_1, N_2, and N_3 are rounded integers. Note that if the desired feed is not reached because of a short tool path, the acceleration and deceleration periods are made equal ($A = D$): i.e. (if $N_2 < 0 \rightarrow N_2 = 0$, $N_1 = N_3 = N/2$).

The interpolation period T_i must be changed at each interval during acceleration and deceleration. Since the traveled tool path segment Δu is kept constant, the following expression can be written between two interpolation periods:

$$\Delta u = \int_{t_{k-1}}^{t_k} Atdt = \frac{A}{2}(t_k^2 - t_{k-1}^2) = \frac{A}{2}(t_k - t_{k-1})(t_k + t_{k-1}).$$

By substituting $T_i(k) = t_k - t_{k-1}$ and $t_k = f(k)/A$, $t_{k-1} = f(k-1)/A$, the interpolation period ($T_i(k)$) during acceleration and deceleration where the velocity changes is found at each increment as

$$T_i(k) = \frac{2\Delta u}{f(k) + f(k-1)}.$$ (5.64)

For small acceleration and deceleration values, the above equation can be approximated as $T_i(k) = \Delta(u)/f(k)$ with caution. The approximation reduces the real time computation load, but it produces jerks at large velocity increments. The following pseudocode algorithm is given to calculate interpolation time intervals for acceleration, constant feed, and deceleration phases:

```
for  k = 1, N₁              ; Iterate during acceleration period.
      f(k) = √(f₀² + 2kAΔu)   ; Next f calculated from initial f₀.
      Tᵢ(k) = 2Δu/(f(k)+f(k-1))  ; Next interpolation period time.
next  k                     ;
```

for $k = 1, N_2$; Iterate during constant feed period.
    $T_i = \Delta u / f$; Period is always the same.
next k ;

for $k = 1, N_3$; Iterate during deceleration period.
    $f(k) = \sqrt{f^2 - 2kD\Delta u}$; Next f calculated from previous f_0.
    $T_i(k) = \frac{2\Delta u}{f(k) + f(k-1)}$; Next interpolation period time.
next k ;

Depending on the *on-the-fly* increase or decrease in the feed, the new values of N_1, N_2, N_3, and T_i are recalculated and updated. Note that acceleration, constant feed, and deceleration phases are coded into different functions and are executed when they are required. As most of the terms are constants, the interpolation rate is constrained by the square root.

Note that the feed, acceleration, and deceleration regulated for vector displacement guarantees synchronized and correct velocity values in all active drives contributing to the vectorial motion in the space. If we take a two-axis motion in the x and y directions, the differential vector displacement Δu can be written as

$$\vec{\Delta u} = \Delta x \vec{i} + \Delta y \vec{j}, \tag{5.65}$$

where i and j are unit vectors in the x and y directions, respectively. If we divide both sides with interpolation time interval T_i, we get

$$\frac{\vec{\Delta u}}{T_i} = \frac{\Delta x}{T_i}\vec{i} + \frac{\Delta y}{T_i}\vec{j},$$
$$\vec{f} = f_x \vec{i} + f_y \vec{j},$$

where the feed amplitude is $f = \sqrt{f_x^2 + f_y^2}$, and f_x and f_y are the resulting velocities of the x and y drives. Hence, once Δu, interpolation time T_i, and counters N_1, N_2, and N_3 are calculated, the velocities and incremental positions in the x and y drives are automatically defined by the algorithm. The interpolation algorithms presented in the next section clarify the engineering implementation of the method.

5.6.2 Jerk-Limited Velocity Profile Generation

The trapezoidal velocity profile introduced in the previous section is simple to implement and suitable for most machines. However, since it employs a constant acceleration, the jerk or the derivative of the acceleration is zero, which leads to various oscillations and noise on the feed and acceleration when interpolating along complex tool paths. The acceleration times inertia or mass acts as a dynamic torque or force on the feed drive structure. If the feed drive acceleration command produced by the trajectory generator is not smooth, the resulting acceleration torque for ball screw and force for linear motor drives

contain high frequency components that excite the structural dynamics of the
feed drives and cause undesired vibrations. To obtain smooth velocity and acceleration profiles, jerk-limited trajectory generation algorithms are used, and
these are presented in this section.

Kinematic Profiles. The kinematic time profiles of jerk (J), acceleration (a),
feed rate (f), and trajectory command position (l) are shown in Figure 5.10.
Before the NC block motion is started, the initial and final values of position
(l_s, l_e) and feed rate (f_s, f_e), maximum acceleration (A), deceleration (D), and
jerk (J) limits are defined. The maximum acceleration/deceleration limits are

Figure 5.10: Kinematic profiles for jerk-limited feedrate generation.

identified from the maximum torque and force limits of the drive motors. The acceleration time is set depending on the peak torque/force delivery periods of the amplifier. The jerk limit is set by the maximum acceleration divided by the acceleration time. From Figure 5.10, the acceleration (a), feed (f), and displacement (l) along the tool path can be expressed as

$$a(t) = a(t_i) + \int_{t_i}^{t} J(\tau)d\tau,$$

$$f(t) = f(t_i) + \int_{t_i}^{t} a(\tau)d\tau, \tag{5.66}$$

$$l(t) = l(t_i) + \int_{t_i}^{t} f(\tau)d\tau.$$

The jerk is constant during acceleration (T_1, T_3) and deceleration (T_5, T_7) periods and is zero during constant acceleration (T_2, T_4, T_6). By integrating the jerk at each period shown in Figure 5.10, the acceleration profiles are obtained:

$$a(\tau) = \begin{cases} J_1\tau_1, & 0 \le t < t_1 \\ A, & t_1 \le t < t_2 \\ A - J_3\tau_3, & t_2 \le t < t_3 \\ 0, & t_3 \le t < t_4 \\ -J_5\tau_5, & t_4 \le t < t_5 \\ -D, & t_5 \le t < t_6 \\ -D + J_7\tau_7, & t_6 \le t < t_7 \end{cases}, \tag{5.67}$$

where τ_k ($k = 1, 2, \ldots, 7$) denotes relative time (i.e., $\tau_k = t - t_{k-1}$). By integrating the acceleration (Eq. 5.67) at each period, the feed velocity profile for each phase in Figure 5.10 is found to be

$$f(\tau) = \begin{cases} f_s + \dfrac{1}{2}J_1\tau_1^2, & f_s : \text{initial feed}, & 0 \le t < t_1 \\[2mm] f_1 + A\tau_2, & f_1 = f_s + \dfrac{1}{2}J_1T_1^2, & t_1 \le t < t_2 \\[2mm] f_2 + A\tau_3 - \dfrac{1}{2}J_3\tau_3^2, & f_2 = f_1 + AT_2, & t_2 \le t < t_3 \\[2mm] f_3, & f = f_3 = f_2 + AT_3 - \dfrac{1}{2}J_3T_3^2, & t_3 \le t < t_4 \\[2mm] f_4 - \dfrac{1}{2}J_5\tau_5^2, & f = f_4 = f_3, & t_4 \le t < t_5 \\[2mm] f_5 - D\tau_6, & f_5 = f_4 - \dfrac{1}{2}J_5T_5^2, & t_5 \le t < t_6 \\[2mm] f_6 - D\tau_7 + \dfrac{1}{2}J_7\tau_7^2, & f_6 = f_5 - DT_6, & t_6 \le t < t_7 \end{cases}, \tag{5.68}$$

where T_k is the period of phase k. Integrating (5.68) once more yields the displacement profile as

$$l(\tau)=\begin{cases} l_{\rm s} + f_{\rm s}\tau_1 + \dfrac{1}{6}J_1\tau_1^3, & l_{\rm s} : \text{initial position}, & 0 \le t < t_1 \\[2mm] l_1 + f_1\tau_2 + \dfrac{1}{2}A\tau_2^2, & l_1 = l_{\rm s} + f_{\rm s}T_1 + \dfrac{1}{6}J_1T_1^3, & t_1 \le t < t_2 \\[2mm] l_2 + f_2\tau_3 + \dfrac{1}{2}A\tau_3^2 - \dfrac{1}{6}J_3\tau_3^3, & l_2 = l_1 + f_1T_2 + \dfrac{1}{2}AT_2^2, & t_2 \le t < t_3 \\[2mm] l_3 + f_3\tau_4, & l_3 = l_2 + f_2T_3 + \dfrac{1}{2}AT_3^2 - \dfrac{1}{6}J_3T_3^3, & t_3 \le t < t_4 \\[2mm] l_4 + f_4\tau_5 - \dfrac{1}{6}J_5\tau_5^3, & l_4 = l_3 + f_3T_4, & t_4 \le t < t_5 \\[2mm] l_5 + f_5\tau_6 - \dfrac{1}{2}D\tau_6^2, & l_5 = l_4 + f_4T_5 - \dfrac{1}{6}J_5T_5^3, & t_5 \le t < t_6 \\[2mm] l_6 + f_6\tau_7 - \dfrac{1}{2}D\tau_7^2 + \dfrac{1}{6}J_7\tau_7^3, & l_6 = l_5 + f_5T_6 - \dfrac{1}{2}DT_6^2, & t_6 \le t < t_7 \end{cases},$$

(5.69)

where l_k is the total displacement reached at the end of phase k. The incremental distance (L_k) travelled at each phase (k) is

$$L_k = l_k - l_{k-1}, \tag{5.70}$$

where the initial displacement is $l_0 = l_{\rm s}$. From the trapezoidal acceleration/ deceleration profiles,

$$A = J_1T_1 = J_3T_3, \quad D = J_5T_5 = J_7T_7 \tag{5.71}$$

should hold, although this may require readjustment after the first initialization step. Considering that the desired feed (f) is reached at the end of phase three, we have

$$f_3 = f \to T_2 = \frac{1}{A}\left[f - f_{\rm s} - \frac{1}{2}J_1T_1^2 - AT_3 + \frac{1}{2}J_3T_3^2\right], \tag{5.72}$$

and similarly, considering that the final feed ($f_{\rm e}$) is reached at the end of phase seven, we have

$$f_7 = f_{\rm e} = f_6 - DT_7 + \frac{1}{2}J_7T_7^2,$$

$$T_6 = \frac{1}{D}\left[f - f_{\rm e} - \frac{1}{2}J_5T_5^2 - DT_7 + \frac{1}{2}J_7T_7^2\right]. \tag{5.73}$$

In addition, the sum of the distances traveled at each step should be equal to the total distance to be traveled (Eqs. 5.69 and 5.70):

$$L = l_{\rm e} - l_{\rm s} = \sum_{k=1}^{7} L_k, \tag{5.74}$$

where L is the total distance traveled within the NC tool path segment.

Initialization. Before generating incremental position commands in the interpolation stage, the number of interpolation steps and the time interval between each of them must be predetermined to implement the jerk-limited kinematic profile given in Figure 5.10. To design a general algorithm, the following conditions are evaluated at the initialization step.

The total number of interpolation steps (N) is checked. If $2 < N \leq 4$, then $N = 4$ is selected to guarantee at least the presence of acceleration and deceleration phases $(1, 3, 5, 7)$ in Figure 5.10. If $N \leq 2$, then $N = 2$ is selected to allow an acceleration and deceleration. Note that these conditions would only occur when the motion is very small, such as in high speed spline interpolation applications or in precision positioning. If the acceleration stage exists, the desired feed (f) must be reached within the first three phases, which implies that $T_2 \geq 0$. If the jerk values are equal $(J_1 = J_3)$ (Eq. 5.71), the acceleration condition requires that

$$T_1 = T_3 = A/J_1, \quad T_2 = \frac{f - f_s}{A} - \frac{A}{J_1} \geq 0. \tag{5.75}$$

If Eq. (5.75) does not hold, than the magnitude of the acceleration must be reduced to its maximum possible limit as

$$A = \text{sgn}(A) \cdot \sqrt{J_1(f - f_s)} \tag{5.76}$$

and T_2 is set to zero. Similarly, if the deceleration stage exists,

$$T_5 = T_7 = D/J_5, \quad T_6 = \frac{f - f_e}{D} - \frac{D}{J_5} \geq 0. \tag{5.77}$$

If the deceleration stage does not exist, the deceleration limit is reduced as

$$D = \text{sgn}(D) \cdot \sqrt{J_5(f - f_e)} \tag{5.78}$$

and T_6 is set to zero. If the displacement length is large enough to encompass the constant feed stage, $T_4 \geq 0$ and must be identified. The total travel length is calculated from Eqs. (5.69) and (5.74) by substituting T_2 and T_6 from Eqs. (5.75) and (5.77) as

$$L = \left(\frac{1}{2A} + \frac{1}{2D}\right) f^2 + \left(\frac{A}{2J_1} + \frac{D}{2J_5} + T_4\right) f + \left(\frac{Af_s}{2J_1} + \frac{Df_e}{2J_5} - \frac{f_s^2}{2A} - \frac{f_e^2}{2D}\right). \tag{5.79}$$

If the tool path distance is too short to reach the target feed (f), its corresponding terms must not be considered in Eq. (5.79). In order to have a constant feed (f) phase, $T_4 \geq 0$ should hold:

$$T_4 = \frac{1}{f}\left[L - \left\{\left(\frac{1}{2A} + \frac{1}{2D}\right) f^2 + \left(\frac{A}{2J_1} + \frac{D}{2J_5}\right) f \right.\right.$$
$$\left.\left. + \left(\frac{Af_s}{2J_1} + \frac{Df_e}{2J_5} - \frac{f_s^2}{2A} - \frac{f_e^2}{2D}\right)\right\}\right] \geq 0. \tag{5.80}$$

If Eq. (5.80) is not satisfied, then $T_4 = 0$ and the magnitude of the feed rate is reduced to its maximum possible value by solving Eq. (5.80). If the quadratic equation yields complex roots, the initial and final feeds are set to zero ($f_s = f_e = 0$), and the path reinitialization is repeated by adjusting the initial parameters (A, D, J).

Number of Interpolation Steps in Each Phase. The incremental displacement Δu is decided by the linear, circular, or spline interpolation algorithms. Since the displacement in each phase is defined in Eq. (5.70), the number of interpolation steps in phases ($1, 3, 5, 7$) can be expressed as

$$N_1 = \text{round}\,(L_1/\Delta u), \quad N_3 = \text{round}\,(L_3/\Delta u),$$
$$N_5 = \text{round}\,(L_5/\Delta u), \quad N_7 = \text{round}\,(L_7/\Delta u). \tag{5.81}$$

If any of (N_1, N_3, N_5, N_7) are zero for nonzero L_1, L_3, L_5, L_7 due to rounded integers, they are set to one. The total number of steps for the acceleration (N_{ac}) and deceleration (N_{dec}) stages are calculated as

$$N_{ac} = \text{round}\,\frac{(L_1 + L_2 + L_3)}{\Delta u}, \quad N_{dec} = \text{round}\,\frac{(L_5 + L_6 + L_7)}{\Delta u}, \tag{5.82}$$

from which the number of steps for constant acceleration (phase 2) and constant deceleration (phase 6) are found:

$$N_2 = N_{ac} - (N_1 + N_3), \quad N_6 = N_{dec} - (N_5 + N_7). \tag{5.83}$$

The number of interpolation steps for the constant feed (phase 4) becomes

$$N_4 = N - (N_{ac} + N_{dec}). \tag{5.84}$$

The number of interpolation steps ($N_1, \ldots, N_7$) for each phase of the jerk-limited velocity profile generation (Fig. 5.10) is now defined. Although the total number of interpolation steps is still N, the displacement at each phase may be changed because of constraints, that is,

$$L_k' = N_k \cdot \Delta u, \quad k = 1, 2, \ldots, 7, \tag{5.85}$$

where L_k' is the new displacement value, which may be different than the value given in Eq. 5.70). This requires readjustment of all acceleration, deceleration, jerk, and time periods again.

For the acceleration stage (phase 1,2,3) when $T_2 > 0$, Eq. (5.72) for T_2 and the new displacements L_1', L_2', L_3' are substituted in Eqs. (5.69) and (5.70), resulting in the following equations:

$$\left. \begin{aligned} &f_s T_1 + \frac{1}{6}AT_1^2 - L_1' = 0 \\[2mm] &-\frac{1}{8}AT_1^2 + \frac{1}{8}AT_3^2 - \frac{1}{2}f_s T_1 - \frac{1}{2}f T_3 + \frac{f^2 - f_s^2}{2A} - L_2' = 0 \\[2mm] &f T_3 - \frac{1}{6}AT_3^2 - L_3' = 0 \end{aligned} \right\} \rightarrow \text{for } T_2 > 0. \tag{5.86}$$

If $T_2 = 0$, Eq. (5.69) can not be used. Instead, the feed rate condition for the end of acceleration stage (f_3 in Eq. 5.68) is used, which leads to

$$
\left.
\begin{aligned}
& f_s T_1 + \frac{1}{6} A T_1^2 - L_1' = 0 \\
& \frac{1}{2} A T_1 + \frac{1}{2} A T_3 + f_s - f = 0 \\
& \frac{1}{3} A T_3^2 + \frac{1}{2} A T_1 T_3 + f_s T_3 - L_3' = 0
\end{aligned}
\right\} \rightarrow \text{ for } T_2 = 0.
\tag{5.87}
$$

A similar approach is applied to the deceleration stage (phases 5,6,7). By substituting T_6 and the new displacements L_5', L_6', L_7' in Eqs. (5.69) and (5.69), we get

$$
\left.
\begin{aligned}
& f T_5 - \frac{1}{6} D T_5^2 - L_5' = 0 \\
& \frac{1}{8} D T_5^2 - \frac{1}{8} D T_7^2 - \frac{1}{2} f T_5 - \frac{1}{2} f_e T_7 + \frac{f^2 - f_e^2}{2D} - L_6' = 0 \\
& f_e T_7 + \frac{1}{6} D T_7^2 - L_7' = 0
\end{aligned}
\right\} \rightarrow \text{ for } T_6 > 0.
$$

$$
\tag{5.88}
$$

For the case $T_6 = 0$ the feed rate condition at the end of deceleration is used:

$$
\left.
\begin{aligned}
& f T_5 - \frac{1}{6} D T_5^2 - L_5' = 0 \\
& \frac{1}{2} D T_5 + \frac{1}{2} D T_7 + f_e - f = 0 \\
& -\frac{1}{3} D T_7^2 - \frac{1}{2} D T_5 T_7 + f T_7 - L_7' = 0
\end{aligned}
\right\} \rightarrow \text{ for } T_6 = 0.
\tag{5.89}
$$

For the acceleration stage, either Eq. (5.86) for $T_2 > 0$ or Eq. (5.87) for $T_2 > 0$ is used to find the new values of T_1, T_3, A, and T_2. For the deceleration stage, either Eq. (5.88) for $T_6 > 0$ or (5.89) for $T_6 = 0$ is used to find the updated values of T_5, T_7, D, and T_6. These equations are nonlinear and can be solved iteratively using the Newton–Raphson numerical algorithm [79]. T_4 is found in Eq. (5.80). Once the values of $T_1, T_2, \ldots, T_7$ and A, D are found, the jerk values are updated using Eqs. (5.71) and (5.77), and the final feed rate values reached at the end of each phase are recalculated from Eq. (5.68).

In short, the number of interpolation steps ($N_1, \ldots, N_7$), time intervals ($T_1, \ldots, T_7$), feed values ($f_1, \ldots, f_7$), displacements ($L_1', \ldots, L_7'$), and possible accelerations and decelerations (A, D, J) are found for each phase of the motion block before the real time interpolation starts.

Recursively Executed Real Time Part. Here the interpolation period at each step must be identified. After the initialization part is completed as explained above, the continuously executed real time part of the jerk-limited feed rate generation algorithm is invoked at each interpolation step. For any of the seven phases in Figure 5.10, the following general displacement formula can be used:

$$l(\tau_k) = \frac{1}{6}J_{0,k}\tau_k^3 + \frac{1}{2}a_{0,k}\tau_k^2 + f_{0,k}\tau_k + l_{0,k}, \tag{5.90}$$

where $J_{0,k}$, $a_{0,k}$, $f_{0,k}$, and $l_{0,k}$ are jerk, acceleration, feed rate, and displacement values respectively, and they are calculated at the initialization stage. τ_k is the relative time parameter at the beginning of each phase k. Since the incremental displacement step Δu is predetermined at the interpolation, the displacement at interpolation step number m can be found as

$$l(\tau_{k,m}) = m \cdot \Delta u = \frac{1}{6}J_{0,k}\tau_{k,m}^3 + \frac{1}{2}a_{0,k}\tau_{k,m}^2 + f_{0,k}\tau_{k,m} + l_{0,k}, \tag{5.91}$$

which leads to the solution of accumulated time $(\tau_{k,m})$ within the phase k. Although the above equation can be solved analytically, it is more efficient to solve it in real time by using the Newton–Raphson iterative algorithm [79]. The interpolation period for within phase k at interpolation step m can be evaluated by

$$T_{i(k,m)} = \tau_{k,m} - \tau_{k,m-1}. \tag{5.92}$$

Alternatively, by replacing the initial conditions at each step, the interpolation period can be solved from Eq. (5.91) as well but at the expense of numerical round-off errors.

Reconstruction of Reference Trajectory at Servo Loop Control Frequency. Reference trajectories generated with varying interpolation periods may cause sudden changes in the discrete position commands at each axis and hence cause undesirable jerks on the drives. To smoothen the generated trajectory, the discrete position commands of each drive at varying interpolation periods must be resampled at the servo loop frequency. Note that the resampling is done after the trajectory is generated by the interpolation stage presented in the following section, although it is presented here because of its relevance to jerk-limited trajectory generation.

A sample trajectory is shown in Figure 5.11, where the interpolator and resampled trajectory at servo control interval T are shown. If we consider one of

Figure 5.11: Resampling of interpolator output Trajectory at Control Loop Frequency.

the drives, for example the x axis, two consecutive displacements of the drive are denoted by x_i, x_{i+1} where i is the interpolation counter. A fifth-order polynomial is fit between the two points as

$$\tilde{x}(\tau) = A_{\mathrm{r}}\tau^5 + B_{\mathrm{r}}\tau^4 + C_{\mathrm{r}}\tau^3 + D_{\mathrm{r}}\tau^2 + E_{\mathrm{r}}\tau + F_{\mathrm{r}}, \tag{5.93}$$

where the time varies as $0 \leq \tau \leq (t_{i+1}-t_i)$. The coefficients $(A_{\mathrm{r}}, B_{\mathrm{r}}, C_{\mathrm{r}}, D_{\mathrm{r}}, E_{\mathrm{r}}, F_{\mathrm{r}})$ are identified from the boundary conditions as

$$
\begin{aligned}
\tilde{x}(0) &= x_i, \quad \tilde{x}(t_{i+1} - t_i) = x_{i+1}, \\
\frac{d\tilde{x}(0)}{d\tau} &= \dot{x}_i, \quad \frac{d\tilde{x}(t_{i+1} - t_i)}{d\tau} = \dot{x}_{i+1}, \\
\frac{d^2\tilde{x}(0)}{d\tau^2} &= \ddot{x}_i, \quad \frac{d^2\tilde{x}(t_{i+1} - t_i)}{d\tau^2} = \ddot{x}_{i+1},
\end{aligned}
\tag{5.94}
$$

where $(\dot{x}_i, \ddot{x}_i; \dot{x}_{i+1}, \ddot{x}_{i+1})$ are the feed rate and acceleration estimates at the beginning and end of the interpolated position command segment, respectively. They can be estimated using a third-order cubic polynomial similar to the method described in the quintic or fifth-order spline interpolation as explained in Section 5.7.3. For each new reference point coming out of the interpolator algorithm, the polynomial coefficients are recalculated, and the reference axis position commands are generated at servo control loop interval T from Eq. (5.93) (i.e., $\tau = 0, T, 2T, \ldots (t_{i+1} - t_i)$).

Example. The trajectory generation with simple trapezoidal velocity and jerk-limited trapezoidal acceleration profiles are compared in Figure 5.12 during a single axis (x) motion. Transient oscillations are present in the acceleration and jerk when the trapezoidal velocity profile is used. However, it can be seen that both acceleration and jerk become very smooth when the trapezoidal acceleration is used instead. The presence of many harmonics in the acceleration indicates that the trapezoidal velocity profile may excite structural modes of the feed drive; hence it is not recommended for high speed machine tool drives. Further smoothing is achieved on the acceleration and jerk when the interpolated axis displacements are resampled using a fifth-order polynomial (see Fig. 5.13).

5.7 REAL TIME INTERPOLATION METHODS

CNC systems must be capable of following the complex trajectories that may be present on parts to be machined. Most of the common geometries can be constructed from linear and circular segments. However, high speed machining of dies, molds, and aerospace parts with sculptured surfaces requires real time spline interpolation as well.

In this section, common design methods for real time two-axis linear, circular, and quintic spline interpolation methods are presented. The algorithms are

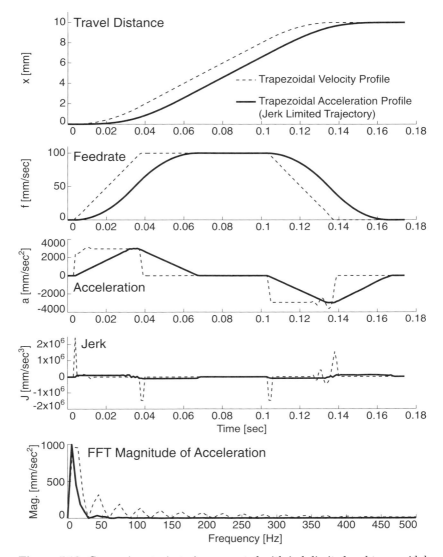

Figure 5.12: Comparison trajectories generated with jerk-limited and trapezoidal velocity profiles.

general and can easily be extended to motions containing more than two active axes [78].

5.7.1   Linear Interpolation Algorithm

This interpolation method is based on the digital integration of the velocity components in two axes.

Let us assume that a cutting tool center is to follow the linear path shown in Figure 5.14. The starting point of the tool is $P_s(x_s, y_s)$ and the end point

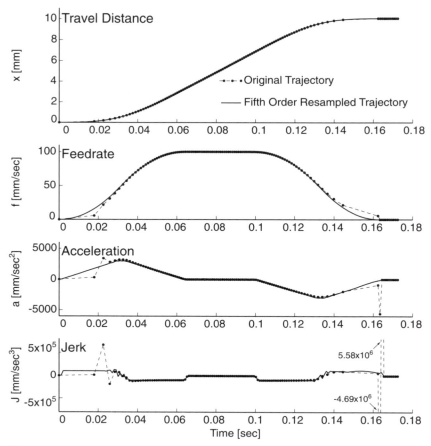

Figure 5.13: Original trajectory generated by the interpolator and smoothed trajectory with 5th order resampling at the axis level.

is $P_e(x_e, y_e)$. At time t the position of the axes will be at

Figure 5.14: Linear interpolation.

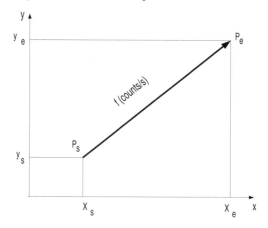

$$x(t) = x(k \cdot \Delta t) = x_s + \int_0^t f_x(t)dt,$$

$$y(t) = y(k \cdot \Delta t) = y_s + \int_0^t f_y(t)dt,$$

$$(5.95)$$

where the axis velocities f_x and f_y and interpolation time interval T_i vary with time during acceleration and deceleration periods but remain constant during the constant feed zone as explained in the previous section. Since the interpolation algorithm is executed N times at time intervals T_i, the above equation (5.95) can be expressed in discrete form as

$$x(k) = x_\text{s} + \sum_{j=1}^{k} f_x(j)T_i(j) = x_\text{s} + \sum_{j=1}^{k-1} f_x(j)T_i(j) + f_x(k)T_i(k),$$

$$y(k) = y_\text{s} + \sum_{j=1}^{k} f_y(j)T_i(j) = y_\text{s} + \sum_{j=1}^{k-1} f_y(j)T_i(j) + f_y(k)T_i(k),$$

or

$$\begin{aligned}
x(k) &= x(k-1) + f_x(k)T_i(k), \\
y(k) &= y(k-1) + f_y(k)T_i(k).
\end{aligned} \tag{5.96}$$

Note that the axis velocities at time interval k are

$$f_x(k) = \frac{\Delta x}{T_i(k)}, \quad f_y(k) = \frac{\Delta y}{T_i(k)}. \tag{5.97}$$

Manipulation of interpolation time interval T_i results in manipulation of the axis feeds according to the vector feed and the acceleration/deceleration velocity profile. However, the incremental displacements in both axes are constant and are given by

$$\Delta x = \frac{x_\text{e} - x_\text{s}}{N}, \quad \Delta y = \frac{y_\text{e} - y_\text{s}}{N}. \tag{5.98}$$

Substituting Eqs. (5.97) and (5.98) into (5.96) gives the recursive digital linear interpolation equations:

$$\begin{aligned}
x(k \cdot \Delta t) &= x_\text{s} + k \cdot \Delta x = x(k-1) + \Delta x, \\
y(k \cdot \Delta t) &= y_\text{s} + k \cdot \Delta y = y(k-1) + \Delta y.
\end{aligned} \tag{5.99}$$

The integration increments Δx and Δy are constant and calculated once at the beginning of the interpolation routine. A real time computer implementation of the linear interpolation is given below.

T_i	The interpolation period.
f	The requested feed rate.
x_s, y_s	Start position.
x_e, y_e	End position.
δx	Total distance in x axis.
δy	Total distance in y axis.
$\text{sign}(x)$	Direction of x movement.
$\text{sign}(y)$	Direction of y movement.
N	The number of interpolation iterations.
dx, dy	Step size for each axis.
x_rem	Remainder for x axis step size.
y_rem	Remainder for y axis step size.

(All values are integer except T_i and f)

The initialization calculations in the CNC are

$$\delta x = \mathrm{abs}(x_e - x_s)$$
$$\delta y = \mathrm{abs}(y_e - y_s)$$
$$\mathrm{sign}(x) = \mathrm{sign}(x_e - x_s)$$
$$\mathrm{sign}(y) = \mathrm{sign}(y_e - y_s)$$
$$\Delta u = f T_i$$
$$N = \sqrt{\delta x^2 + \delta y^2}/\Delta u$$
$$dx = \mathrm{fix}(\delta x/N)$$
$$dy = \mathrm{fix}(\delta y/N)$$
$$x_{\mathrm{rem}} = \delta x - (dx \cdot N)$$
$$y_{\mathrm{rem}} = \delta y - (dy \cdot N)$$
$$\mathrm{line}(x_s, dx, x_{\mathrm{rem}}, \mathrm{sign}(x), N) \qquad \text{; sent to } x \text{ axis position controller}$$
$$\mathrm{line}(y_s, dy, y_{\mathrm{rem}}, \mathrm{sign}(y), N) \qquad \text{; sent to } y \text{ axis position controller}$$

After initialization of as many axes as are needed, each can be calculated at every period T_i with the following:

```
function line (x_s, dx, x_rem, sign(x), N)
x(1) = x_s
x_error = 0
for i = 2, N + 1
    x(i) = x(i − 1) + sign(x) · dx
    x_error = x_error + x_rem
    if (x_error >= N)
        x(i) = x(i) + sign(x)
        x_error = x_error − N
    end
end
```

The generated position commands at each interpolation time interval T_i are then sent to the digital servo control algorithm of each axis in the CNC.

Example. The following NC block is entered to the CNC in *counts*:

N010 G01 G90 X24 Y32 F1000

This NC program provides $P_e(24, 32)$ and $f = 1000$ [counts/s]. The initial coordinates were given as $P_s(5, 6)$, the acceleration and deceleration values of the CNC are set as $A = D = 50{,}000$ [counts/s²], and the trapezoidal velocity profile is used. The minimum interpolation time is $T_{\min} = 0.002\,\mathrm{s}$ in the CNC. The interpolation algorithm is set as follows:

$$\delta x = x_e - x_s = 19, \quad \delta y = y_e - y_s = 26, \quad L = \sqrt{\delta x^2 + \delta y^2} = 32,$$
$$\Delta u = f T_{\min} = 2, \quad N = L/\Delta u = 16, \quad N_1 = N_3 = f^2/(2A\Delta u) = 5,$$
$$\Delta x = \delta x/N = 1.1875, \quad \Delta y = \delta y/N = 1.625.$$

The numerical interpolation results are shown in Table 5.3 and Figure 5.15.

TABLE 5.3. Linear Interpolation Steps between Points $P_s(5, 6)$ and $P_e(24, 32)$

Clock Pulses k	Feed $f(k)$ [counts/s]	Interpolation Interval T_i (ms)	Time t [ms]	$x(k)$ [counts]	$y(k)$ [counts]
0	0.00	0.00000	0.0	5.00	6.00
1	447.21	0.00894	8.9	6.19	7.63
2	632.46	0.00370	12.6	7.38	9.25
3	774.60	0.00284	15.5	8.56	10.88
4	894.43	0.00240	17.9	9.75	12.50
5	1000.00	0.00211	20.0	10.94	14.13
6	1000.00	0.00200	22.0	12.13	15.75
7	1000.00	0.00200	24.0	13.31	17.38
8	1000.00	0.00200	26.0	14.50	19.00
9	1000.00	0.00200	28.0	15.69	20.63
10	1000.00	0.00200	30.0	16.88	22.25
11	1000.00	0.00200	32.0	18.06	23.88
12	894.43	0.00211	34.1	19.25	25.50
13	774.60	0.00240	36.5	20.44	27.13
14	632.46	0.00284	39.4	21.63	28.75
15	447.21	0.00370	43.1	22.81	30.38
16	0.00	0.00894	52.0	24.00	32.00

5.7.2   Circular Interpolation Algorithm

Consider a circular segment whose center is at the origin of the CNC's coordinate system (see Fig. 5.16). The circular tool path has a length of

$$L = R(\theta_e - \theta_s) = R\theta_t. \tag{5.100}$$

The arc must be broken into N small segments for digital interpolation. The length of each segment is Δu, with a corresponding angular segment of $\Delta\theta$ as shown in Figure 5.16. The chord error (chord_error) must be kept less than the resolution of position sensing systems (i.e., 1 count):

$$\text{chord\_error} = R\left(1 - \cos\frac{\Delta\theta}{2}\right) \leq 1, \tag{5.101}$$

which is satisfied when the angular segment $\Delta\theta$ is selected as [66]

$$\Delta\theta \leq 2\cos^{-1}\left(\frac{R-1}{R}\right). \tag{5.102}$$

The selection of

$$\Delta\theta = \cos^{-1}\left(\frac{R-1}{R}\right), \quad \cos\Delta\theta = \frac{R-1}{R} \tag{5.103}$$

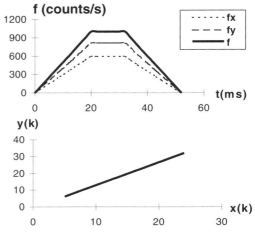

Figure 5.15: Linear interpolation example.

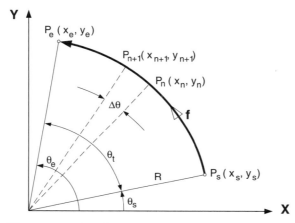

Figure 5.16: Circular interpolation.

guarantees half a count chord error and is more advantageous for computation. The corresponding chord segment is found as

$$\Delta u = R\Delta\theta. \qquad (5.104)$$

The circular tool path is broken into N_1, N_2, N_3 segments for trapezoidal velocity profiling as explained earlier. The tool has a feed speed of f [counts/s] on the arc, and the arc segment Δu is traveled in one interpolation period T_i. The interpolation period T_i varies during acceleration and deceleration zones (N_1, N_3), but it remains constant during the steady state feed zone (N_2) as shown in Figure 5.9. The angular feed velocity $\omega(t)$ and instantaneous angular position $(\theta(t))$ of the cutter at time t are

$$\omega = \frac{f}{R}, \quad \theta(t) = \omega t = \frac{f}{R}t \quad (5.105)$$

respectively. Coordinates of a point on the arc can be expressed as

$$x(t) = R \cos\theta(t) = R \cos\left(\frac{f}{R}t\right),$$

$$y(t) = R \sin\theta(t) = R \sin\left(\frac{f}{R}t\right).$$

$$(5.106)$$

The velocities of the x and y axis feed drives are

$$f_x = \frac{dx}{dt} = -\frac{f}{R}R\sin\left(\frac{f}{R}t\right) = -\frac{f}{R} \cdot y(t),$$

$$f_y = \frac{dy}{dt} = \frac{f}{R}R\cos\left(\frac{f}{R}t\right) = \frac{f}{R} \cdot x(t).$$

$$(5.107)$$

Because the velocities are coupled with the positions, digital integration of the above equations always produces some errors. A decoupled recursive circular interpolation explained in the following can be easily implemented on CNC systems that are capable of handling floating point arithmetic operations [80].

Consider two points, P_n and P_{n+1} on the arc, which are $\Delta\theta$ away from each other (Figure 5.16):

$$x_n = R \cos(\theta_s + n\Delta\theta), \quad y_n = R \sin(\theta_s + n\Delta\theta),$$
$$x_{n+1} = R \cos[\theta_s + (n+1)\Delta\theta], \quad y_{n+1} = R \sin[\theta_s + (n+1)\Delta\theta]. \tag{5.108}$$

The coordinates of P_{n+1} can be expanded as

$$x_{n+1} = R \cos(\theta_s + n\Delta\theta + \Delta\theta),$$
$$y_{n+1} = R \sin(\theta_s + n\Delta\theta + \Delta\theta),$$

or

$$x_{n+1} = R\cos(\theta_s + n\Delta\theta)\cos\Delta\theta - R\sin(\theta_s + n\Delta\theta)\sin\Delta\theta,$$
$$y_{n+1} = R\sin(\theta_s + n\Delta\theta)\cos\Delta\theta + R\cos(\theta_s + n\Delta\theta)\sin\Delta\theta.$$

The products of trigonometric functions can be expanded as

$$\sin(\theta_s + n\Delta\theta)\sin\Delta\theta = \frac{1}{2}\{\cos[\theta_s + (n-1)\Delta\theta] - \cos[\theta_s + (n+1)\Delta\theta]\},$$

$$\cos(\theta_s + n\Delta\theta)\cos\Delta\theta = \frac{1}{2}\{-\sin[\theta_s + (n-1)\Delta\theta] + \sin[\theta_s + (n+1)\Delta\theta]\}.$$

Substituting the trigonometric expansion into the coordinates of P_{n+1}, we get

$$x_{n+1} = R \cos(\theta_s + n\Delta\theta)\cos\Delta\theta - \frac{R}{2}\cos[\theta_s + (n-1)\Delta\theta] + \frac{R}{2}\cos[\theta_s + (n+1)\Delta\theta],$$

$$y_{n+1} = R \sin(\theta_s + n\Delta\theta)\cos\Delta\theta - \frac{R}{2}\sin[\theta_s + (n-1)\Delta\theta] + \frac{R}{2}\sin[\theta_s + (n+1)\Delta\theta].$$

When the coordinates of x_n and y_n are substituted in Eq. (5.108), the following difference equations are obtained:

$$x_{n+1} = x_n \cos\Delta\theta - \frac{1}{2}x_{n-1} + \frac{1}{2}x_{n+1},$$

$$y_{n+1} = y_n \cos\Delta\theta - \frac{1}{2}y_{n-1} + \frac{1}{2}y_{n+1}.$$

The discrete position equations can be reduced to a set of decoupled recursive axis equations:

$$x_{n+1} = 2x_n \cos\Delta\theta - x_{n-1},$$
$$y_{n+1} = 2y_n \cos\Delta\theta - y_{n-1}. \tag{5.109}$$

Note that the integration interval $\Delta\theta$ is selected according to the maximum chord error criterion, and $\cos\Delta\theta$ is given by Eq. (5.103). The interpolation time intervals are calculated according to the velocity profile as explained in Section 5.6. The numerical value of $\Delta\theta$ is precomputed and stored in the memory before the real time, recursive interpolation algorithm starts.

The algorithm has been tested on the research milling machine controlled by an in-house developed CNC system [81]. A two-axis 117 count (0.148 mm) linear move, followed by an 80 count (0.102 mm) move in the x direction and a full circular path with 80 count (0.102 mm) radius were executed on the machine tool at an intended feed of 4,000 counts/s. The command and measured tool paths, actual feed along the tool path, and the corresponding axis velocities are shown in Figure 5.17. With $A = 50,000$ counts/s^2 acceleration, the target feed is

Figure 5.17: A sample tool path with combined linear and circular interpolation segments. The actual position and velocities were evaluated from encoder measurements. 1 count = 0.00127 mm.

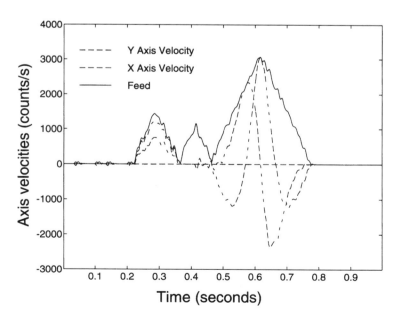

never reached because of the short travel distances. However, the path is still followed with a satisfactory accuracy.

Example. The following NC block is entered to the CNC in counts:

N010 G90 G03 X90 Y-90 I-90 J0 F1000

This NC program provides $P_e(90, -90)$, $f = 1,000$ [counts/s], and $R = \sqrt{I^2 + J^2} = 90$ [counts]. The initial coordinates were given as $P_s(180, 0)$, and the acceleration (A) and deceleration (D) values of the CNC are set to $A = D = 4,000$ [counts/s^2] with a trapezoidal velocity profile. The minimum interpolation time is $T_{\min} = 0.002$ s in the CNC. The interpolation algorithm is set as follows:

$$\theta_s = 0, \quad \theta_e = \theta_t = 3\pi/2 = 4.71239, \quad \Delta\theta = 0.298419,$$
$$\Delta u = R\Delta\theta = 26.857723, \quad N_1 = N_3 = f^2/(2A\Delta u) = 5, \quad N = \theta_t/\Delta\theta = 16.$$

The numerical interpolation results are shown in Table 5.4 and Figure 5.18.

5.7.3 Quintic Spline Interpolation within CNC Systems

In addition to linear and circular interpolation, modern CNCs offer parabolic, helical, and spline interpolations. Instead of breaking sculptured paths into small linear and circular segments, it is preferred to move the cutting tool along the complex path using the path geometry itself [82]. This reduces the length of the NC program and improves the smooth generation of velocity and acceleration in addition to the contouring along the path. Real time quintic spline interpolation is most useful in high speed machining of dies and molds [83], and its implementation in the CNC is presented here.

TABLE 5.4. Circular Interpolation Steps between Points $P_s(180, 0)$ and $P_e(90, -90)$

Clock Pulses k	Feed $f(k)$ [counts/s]	Interpolation Interval T_i (ms)	Time t [ms]	$x(k)$ [counts]	$y(k)$ [counts]
0	0.00	0.0000	0	180.00	0.00
1	463.53	0.1158	116	176.02	26.46
2	655.53	0.0480	164	164.44	50.58
3	802.86	0.0368	201	146.28	70.23
4	927.06	0.0310	232	123.14	83.68
5	1000.0	0.0278	260	97.08	89.72
6	1000.0	0.0268	286	70.38	87.84
7	1000.0	0.0268	313	45.43	78.19
8	1000.0	0.0268	340	24.41	61.63
9	1000.0	0.0268	367	9.19	39.62
10	1000.0	0.0268	394	1.11	14.11
11	1000.0	0.0268	421	0.89	−12.65
12	886.08	0.0284	449	8.55	−38.29
13	755.17	0.0327	482	23.41	−60.54
14	596.17	0.0397	522	44.15	−77.45
15	374.90	0.0553	577	68.95	−87.50
16	0.00	0.1432	720	90.00	−90.00

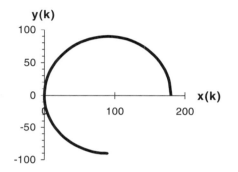

Figure 5.18: Circular interpolation example.

Figure 5.19: Spline passing through knots and the use of cubic polynomials in estimating the first and second derivatives.

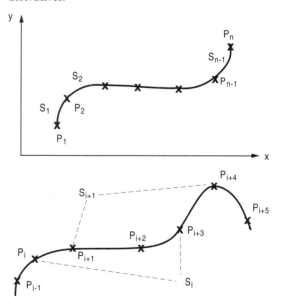

A series of n knots $(P_1, P_2, \ldots, P_n)$ along the tool path are to be connected by quintic spline segments with smooth transitions (see Fig. 5.19). A spline segment S_i connecting two knots (P_i, P_{i+1}) is expressed by a fifth-order polynomial as follows:

$$\{S_{qi}\} = \{A_{qi}\}u^5 + \{B_{qi}\}u^4 + \{C_{qi}\}u^3 + \{D_{qi}\}u^2 + \{E_{qi}\}u + \{F_{qi}\},$$

(5.110)

where $q : \{x, y, z\}, u \in [0, l]$ and $\{A_{qi}\}$, $\{B_{qi}\}, \{D_{qi}\}, \{E_{qi}\}, \{F_{qi}\}$ are vectors containing the coefficients for each coordinate (i.e., $q : x, y, z$ for a three-axis machine), and u is the spline parameter that changes between zero and spline segment length l_i. The number of elements in each vector is therefore equal to the number of axes in motion. The real time interpolation algorithm requires the values of all spline coefficients (A, B, C, D, E, F), the spline length (l_i), and the feed, acceleration, and jerk limits of the machine tool drives as input parameters. The quintic spline interpolation has both off-line and on-line mathematical steps. First, a quintic spline is fit to a series of knots along a sculptured tool path as explained in Section 4.7, and the coefficients are identified. The fitting can either be done on the CAD/CAM system during tool path generation, and the coefficients of the spline can be transferred to CNC via NC codes, or they can be identified within the preprocessing stage of CNC by using the raw knot coordinates, supplied within the NC program. The length of the spline segment is estimated by approximating its chord length. The coefficients, knot coordinates, and approximated chord length are used to generate individual axis coordinates in real time by considering the feed, acceleration, and jerk limits of the machine tool drives.

Each step in the implementation is shown in Figure 5.20 and explained in the following sections.

Evaluation of Quintic Spline Parameters

Since there are six coefficients, we need six boundary conditions to evaluate them. These are the coordinates (P_i, P_{i+1}), first derivatives $(\frac{dP_i}{du}, \frac{dP_{i+1}}{du})$, and the second derivatives $(\frac{d^2P_i}{du^2}, \frac{d^2P_{i+1}}{du^2})$ at the two knots (P_i, P_{i+1}). The first and second derivatives of the quintic spline (5.110) with respect to u are

$$\left\{\frac{dS_i}{du}\right\} = 5\{A_i\}u^4 + 4\{B_i\}u^3$$
$$+ 3\{C_i\}u^2 + 2\{D_i\}u + \{E_i\},$$

$$\left\{\frac{d^2S_i}{du^2}\right\} = 20\{A_i\}u^3 + 12\{B_i\}u^2$$
$$+ 6\{C_i\}u + 2\{D_i\}. \qquad (5.111)$$

The coefficients are evaluated from the following matrix:

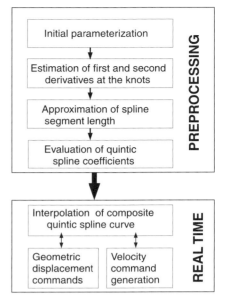

Figure 5.20: Preprocessing and real time processing steps of quintic spline interpolation.

$$\begin{bmatrix} u_0^5 & u_0^4 & u_0^3 & u_0^2 & u_0 & 1 \\ u_1^5 & u_1^4 & u_1^3 & u_1^2 & u_1 & 1 \\ 5u_0^4 & 4u_0^3 & 3u_0^2 & 2u_0 & 1 & 0 \\ 5u_1^4 & 4u_1^3 & 3u_1^2 & 2u_1 & 1 & 0 \\ 20u_0^3 & 12u_0^2 & 6u_0 & 2 & 0 & 0 \\ 20u_1^3 & 12u_1^2 & 6u_1 & 2 & 0 & 0 \end{bmatrix} \begin{Bmatrix} A_i \\ B_i \\ C_i \\ D_i \\ E_i \\ F_i \end{Bmatrix} = \begin{Bmatrix} P_i \\ P_{i+1} \\ \dfrac{dP_i}{du} \\ \dfrac{dP_{i+1}}{du} \\ \dfrac{d^2P_i}{du^2} \\ \dfrac{d^2P_{i+1}}{du^2} \end{Bmatrix}, \qquad (5.112)$$

which leads to the solution of six quintic spline coefficients with $u_0 = 0, u_1 = l_i$:

$$\{A_i\} = \frac{1}{u_1^5}\left[6(P_{i+1} - P_i) - 3\left(\frac{dP_{i+1}}{du} + \frac{dP_i}{du}\right)u_1 + \frac{1}{2}\left(\frac{d^2P_{i+1}}{du^2} - \frac{d^2P_i}{du^2}\right)u_1^2\right],$$

$$\{B_i\} = \frac{1}{u_1^4}\left[-15(P_{i+1} - P_i) + \left(7\frac{dP_{i+1}}{du} + 8\frac{dP_i}{du}\right)u_1 - \left(\frac{d^2P_{i+1}}{du^2} - \frac{3}{2}\frac{d^2P_i}{du^2}\right)u_1^2\right],$$

$$\{C_i\} = \frac{1}{2u_1^3}\left[20(P_{i+1} - P_i) - 4\left(2\frac{dP_{i+1}}{du} + 3\frac{dP_i}{du}\right)u_1 + \left(\frac{d^2P_{i+1}}{du^2} - 3\frac{d^2P_i}{du^2}\right)u_1^2\right],$$

$$\{D_i\} = \frac{1}{2}\frac{d^2P_i}{du^2},$$

$$\{E_i\} = \frac{dP_i}{du},$$

$$\{F_i\} = P_i. \qquad (5.113)$$

In order to evaluate the coefficients (5.113), the first and second derivatives at the two ends (P_i, P_{i+1}) of the spline segment must be known.

Identification of Derivatives Using Parabolic Approximation

The derivatives at the knots are estimated by approximating a parabola between three consecutive knots. Consider the following parabola:

$$\{S_i(u)\} = \{a_i\} u^2 + \{b_i\} u + \{c_i\}, \quad u \in [0, L_i, L_{i+1}]. \tag{5.114}$$

passing through three consecutive knots P_i, P_{i+1}, P_{i+2}, that is,

$$\begin{bmatrix} u_0^2 & u_0 & 1 \\ u_1^2 & u_1 & 1 \\ u_2^2 & u_2 & 1 \end{bmatrix} \begin{Bmatrix} \{a_i\} \\ \{b_i\} \\ \{c_i\} \end{Bmatrix} = \begin{Bmatrix} \{P_i\} \\ \{P_{i+1}\} \\ \{P_{i+2}\} \end{Bmatrix}, \tag{5.115}$$

where $u_0 = 0$, $u_1 = L_i$, and $u_2 = L_i + L_{i+1}$. The distance between each knot is approximated as a line:

$$L_i = |P_{i+1} - P_i| = \sqrt{(x_{i+1} - x_i)^2 + (y_{i+1} - y_i)^2 + (z_{i+1} - z_i)^2} \tag{5.116}$$

for a three-axis path. The parameters of the parabola can be identified from the matrix expression as Eq. (5.115) as

$$\{a_i\} = \frac{(P_{i+2} - P_i)(u_1 - u_0) - (P_{i+1} - P_i)(u_2 - u_0)}{(u_2^2 - u_0^2)(u_1 - u_0) - (u_1^2 - u_0^2)(u_2 - u_0)},$$

$$\{b_i\} = \frac{(P_{i+1} - P_i) - (u_1^2 - u_0^2)\{a_i\}}{(u_1 - u_0)},$$

$$\{c_i\} = P_i - u_0^2\{a_i\} - u_0\{b_i\}. \tag{5.117}$$

The first and second derivatives of the parabola at knot P_i are

$$\left\{ \frac{dS_i(u)}{du} \right\} = 2\{a_i\}u + \{b_i\},$$

$$\left\{ \frac{d^2 S_i(u)}{du^2} \right\} = 2\{a_i\}. \tag{5.118}$$

Note that the second derivative has a unique solution, and the first derivative has three solutions using $u_0 = 0$, $u_1 = L_i$, and $u_2 = L_i + L_{i+1}$ at subsequent knots. An average value of the first derivative can be used for better accuracy at the mid knots. The solution is unique at the first and last knot.

Identification of Derivatives Using Cubic Spline Approximation

Experience has indicated that the parabolic approximation of the spline with a linear chord length assumption does not lead to smooth feed rate changes along the quintic spline path. Although computationally more demanding, a cubic spline approximation for the derivatives at the knots and a near–arc length estimation for the chord length between the knots are found to generate smoother feed and acceleration during quintic spline interpolation.

Consider a cubic spline passing through the knots P_i, P_{i+1}, P_{i+2}, P_{i+3} shown in Figure 5.19:

$$\{S_i(u)\} = \{a_i\}\,u^3 + \{b_i\}\,u^2 + \{c_i\}\,u + \{d_i\}, \quad u \in [0, L_i + L_{i+1} + L_{i+2}], \qquad (5.119)$$

where L_i is the length of the chord segment between knots (P_i, P_{i+1}) and is to be estimated using near–arc length parameterization. The number of elements in each vector is equal to the number of active axes on the machine. The boundary conditions at each knot are

$$\{S_i(u)\} = \begin{cases} P_i, & u_0 = 0, \\ P_{i+1}, & u_1 = L_i, \\ P_{i+2}, & u_2 = L_i + L_{i+1}, \\ P_{i+3}, & u_3 = L_i + L_{i+1} + L_{i+2}. \end{cases} \qquad (5.120)$$

By substituting the boundary conditions at four knots (5.120) into the cubic spline equation (5.119) the spline parameters can be evaluated:

$$\begin{bmatrix} u_0^3 & u_0^2 & u_0 & 1 \\ u_1^3 & u_1^2 & u_1 & 1 \\ u_2^3 & u_2^3 & u_2^3 & 1 \\ u_3^3 & u_3^3 & u_3^3 & 1 \end{bmatrix} \begin{Bmatrix} \{a_i\} \\ \{b_i\} \\ \{c_i\} \\ \{d_i\} \end{Bmatrix} = \begin{Bmatrix} P_i \\ P_{i+1} \\ P_{i+2} \\ P_{i+3} \end{Bmatrix}, \qquad (5.121)$$

which leads to the following solution:

$$\{a_i\} = \left[\frac{(P_{i+1} - P_i)u_2 u_3 (u_2 - u_3) + (P_{i+2} - P_i)u_3 u_1 (u_3 - u_1) + (P_{i+3} - P_i)u_1 u_2 (u_1 - u_2)}{\Delta_c} \right],$$

$$\{b_i\} = \left[\frac{(P_{i+1} - P_i)u_2 u_3 \left(u_3^2 - u_2^2\right) + (P_{i+2} - P_i)u_3 u_1 \left(u_1^2 - u_3^2\right) + (P_{i+3} - P_i)u_1 u_2 \left(u_2^2 - u_1^2\right)}{\Delta_c} \right],$$

$$\{c_i\} = \left[\frac{(P_{i+1} - P_i)u_2^2 u_3^2 (u_2 - u_3) + (P_{i+2} - P_i)u_3^2 u_1^2 (u_3 - u_1) + (P_{i+3} - P_i)u_1^2 u_2^2 (u_1 - u_2)}{\Delta_c} \right],$$

$$\{d_i\} = \{P_i\}, \qquad (5.122)$$

where

$$\Delta_c = u_1 u_2 u_3 \left[u_1^2(u_2 - u_3) + u_2^2(u_3 - u_1) + u_3^2(u_1 - u_2)\right].$$

The derivatives of the spline at each knot are

$$\left. \begin{aligned} \frac{dS_i}{du} &= 3\{a_i\}u^2 + 2\{b_i\}u + \{c_i\} \\ \frac{d^2 S_i}{du^2} &= 6\{a_i\}u + 2\{b_i\} \end{aligned} \right\} \quad u \in [0, \, u_3]. \qquad (5.123)$$

If there are n knots on the spline, the derivatives at the first two knots (P_1, P_2) and the last two knots are evaluated using the coordinates of the first four (P_1, P_2, P_3, P_4) and last four $(P_{n-3}, P_{n-2}, P_{n-1}, P_n)$. The derivatives at the remaining knots are evaluated for two mid knots of each spline segment that spans over four points. For example, spline segment S_i is used to evaluate the derivatives at knots P_{i+1}, P_{i+2} in Figure 5.19. This will lead to two values for each derivative, which are obtained from two consecutive spline segments. An average of the two derivative values is used in the quintic spline equation (5.113).

Near–Arc Length Parameterization

The length of a spline segment between the two consecutive knots (P_i, P_{i+1}) is better approximated by assuming an arc length instead of a straight line. When the spline (dS_i) is parameterized by its arc length, it must have a unit tangent along the spline segment. A closed arc length can be found when the deviation of (dS/du) is minimized. The least square solution is formulated as

$$\frac{d}{du} \int_0^l \left(\frac{dS}{du} - 1 \right)^2 du = 0. \tag{5.124}$$

An analytical solution of the length (l) from this equation is difficult. A reasonable solution can be obtained by considering the midpoint of the chord;

$$\left| \frac{dS}{du} \left(u = \frac{l}{2} \right) \right| = 1. \tag{5.125}$$

By substituting $u = l/2$ and quintic spline coefficients (5.113) into the first derivative of quintic spline (Eq. 5.111), we get

$$\left| \frac{15}{8l_i}(P_{i+1} - P_i) - \frac{7}{16} \left(\frac{dP_{i+1}}{du} + \frac{dP_i}{du} \right) + \frac{1}{32} \left(\frac{d^2 P_{i+1}}{du^2} - \frac{d^2 P_i}{du^2} \right) l_i \right| = 1. \tag{5.126}$$

Squaring both sides of the equation leads to

$$\left(\frac{15}{8l_i} \Delta P \right)^2 + \left(\frac{-7}{16} \left(\frac{dP_{i+1}}{du} + \frac{dP_i}{du} \right) \right)^2 + \left(\frac{l_i}{32} \Delta^2 P \right)^2$$
$$- \frac{105}{64 l_i} \Delta P \left(\frac{dP_{i+1}}{du} + \frac{dP_i}{du} \right)^2 - \frac{7}{256} \Delta^2 P \left(\frac{dP_{i+1}}{du} + \frac{dP_i}{du} \right) l_i$$
$$+ \frac{15}{128} \Delta P \cdot \Delta^2 P = 1, \tag{5.127}$$

where $\Delta P = P_{i+1} - P_i$ and $\Delta^2 P = \frac{d^2 P_{i+1}}{d^2 u} - \frac{d^2 P_i}{d^2 u}$. Equation (5.127) is rearranged as a fourth-order polynomial,

$$al_i^4 + bl_i^3 + cl_i^2 + dl_i + e = 0, \tag{5.128}$$

where the coefficients are

$$a = \left(\frac{d^2 P_{i+1}}{d^2 u} - \frac{d^2 P_i}{d^2 u} \right)^2,$$

$$b = -28 \left(\frac{d^2 P_{i+1}}{d^2 u} - \frac{d^2 P_i}{d^2 u} \right) \left(\frac{dP_{i+1}}{du} + \frac{dP_i}{du} \right),$$

$$c = -1024 + 196 \left(\frac{dP_{i+1}}{du} + \frac{dP_i}{du} \right)^2 + 120 \left(P_{i+1} - P_i \right) \left(\frac{d^2 P_{i+1}}{d^2 u} - \frac{d^2 P_i}{d^2 u} \right),$$

$$d = -1680 \left(P_{i+1} - P_i \right) \left(\frac{dP_{i+1}}{du} + \frac{dP_i}{du} \right),$$

$$e = 3600 \left(P_{i+1} - P_i \right)^2 .$$

Equation (5.128) can be solved by an iteration scheme such as the Newton–Raphson method [79]. Estimation of spline segment length (l_i) by using the near–arc length approximation (Eq. 5.128) minimizes the deviation of the unit tangent vector much better than the linear chord length parameterization; hence it leads to a smoother feed rate and acceleration during quintic spline interpolation.

Spline interpolation is summarized as follows (Fig. 5.20):

Estimate the derivatives at the knots from Eq. (5.123).

Estimate the chord length (l_i) from Eq. (5.128).

Evaluate the quintic spline coefficients from Eq. (5.113) by substituting the chord length (l_i) and derivatives.

Generate incremental motion commands from the quintic spline expression (5.110) by varying $u = 0, \delta u, 2\delta u, \ldots, k\delta u, \ldots, N\delta u$, where $\delta u = l_i / N$. Adjust the number of interpolation steps and the time intervals depending on the feed, acceleration, and position feedback measurement resolution as explained in the linear and circular interpolation algorithms.

Examples. The quintic spline interpolation and jerk-limited trajectory generation are implemented within an open CNC developed in house [76]. A spiral tool path has been generated in real time on a two-axis high speed XY table. The actual spiral is generated by a pen attached to the stationary spindle, and resulting feed, acceleration, and jerk on each drive and along the tool path are shown in Figure 5.21. The beginning of the tool path corresponds to the center of the spiral, where some transients can be seen on the acceleration and jerk. A second example is shown in Figure 5.22, where a complex sculptured surface was represented with a series of quintic splines in a preprocessing stage. The part is machined on a three-axis milling machine controlled by the open CNC. The machined part is also shown in Figure 5.22. Both examples employ a quintic spline with jerk-limited trajectory generation.

5.8 DESIGN OF AN ELECTRO-HYDRAULIC CNC PRESS BRAKE

Sheet metal–forming machine tools, *press brakes*, are widely used in fabrication shops. A typical press brake consists of a moving ram, which holds a punch, and a die located on a bed frame (see Fig. 5.23). The motion of the ram is delivered by a pair of hydraulic actuators. Conventional presses are manually operated, and the end positions of the ram are manually set via limit switches or dead stops. Manual control leads to frequent tuning and adjustment of the machine and requires a number of trial bends until a satisfactory positioning and bending accuracy are achieved. In contrast the CNC press provides flexibility in adapting the machine to different bending operations quickly, while delivering an accurate positioning. A simplified design of a computer control system for a press brake is presented in this section.

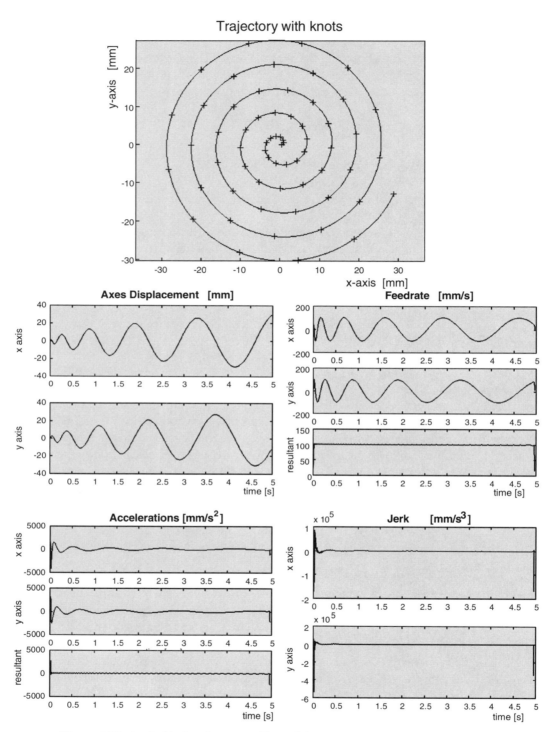

Figure 5.21: A spiral tool path generated by real time quintic spline interpolation and corresponding feedrate, acceleration, and jerk values produced by jerk-limited trajectory generation.

5.8.1 Hydraulic Press Brake System

Sample tool trajectory

Mechanical System. The press ram, which holds the punch at its free end, is held by two parallel hydraulic actuators (Fig. 5.23). The ram slides on lubricated guides, which are modified to have cylindrical pads. The pads allow 3° tilt to the ram–punch assembly for press bending of sheet metals with a variable penetration depth at either side. The aim of the CNC system is to bring the punch rapidly toward the sheet metal laid on the V-shaped die and to form it at high pressure but with a slow penetration velocity to a desired depth. Depending on the offset between the two ends of the bend, the positions and velocities of the two hydraulic actuators that are holding the punch must be controlled by the CNC system.

The actual position of each side of the ram is monitored by linear encoders. In order to consider the relative static deformations between the press bed and ram, a sliding tie-rod is designed as shown in Figure 5.23. The tie-rod posts are placed at the axes of the ram/actuator connections. High precision ball joints are used to connect each tie-rod to its respective post and slider. A linear encoder with 0.005 mm resolution was selected. The stationary encoder body was mounted on the press bed close to the die, and the moving read head was attached to the linear slider housed in a cylindrical guide with an adjustable stiffness. In place of the encoder, precision Linear Displacement Transducers (LDTs), which use a tubular wave guide and a magnet, can be used as an integral part of the actuator rods.

Figure 5.22: A sample sculptured surface tool path and machined part using quintic spline interpolator and jerk-limited trajectory generation.

Hydraulic System. The schematic diagram of the hydraulic system design is shown in Figure 5.23. A variable displacement pump with pressure feedback was chosen to deliver the hydraulic power. An accumulator, which keeps the system pressure within 10% of a set level and damps the pressure oscillations caused by the reciprocating pump pistons, was used. A check-valve was placed between the accumulator and the pump to eliminate the possibility of the accumulator pressure driving the pump in reverse in the event of power outage. A counter-balance valve was used to lock the system whenever the system pressure is lost. A fine pressure filter was used in the supply line to keep any debris from reaching the sensitive pilot stage of the servovalves. A coarse low pressure filter was added to the return line. The existing hydraulic actuators were modified to include servovalves and pressure transducers as shown in Figure 5.24, which would have been an integral part of the actuators on a production machine. The flow into and out of the actuator cylinder is controlled by a two-stage servovalve. It is rated as having a 20 gallon/min flow capacity with 85 Hz

Press Brake

Sheet metal bending operation

Hydraulic circuit of the press brake

Figure 5.23: Hydraulic press brake system.

bandwidth by the manufacturer. The primary stage consists of an electronic torque motor driving the primary flapper valve, which provides a differential pressure across the ends of the secondary closed center spool valve with a zero overlap. The displacement of the secondary spool is fed back to the torque motor by way of a cantilever spring. This spool controls the flow into the cylinder.

The hydraulic cylinder has two fluid chambers separated by a piston, which is connected to the ram from one side with a rigid rod. The inertia loads of the rod and ram assembly, coulomb and viscous friction loads in the ram guideways and piston cylinder interface, and the bending loads must be overcome by the pressurized fluid injected into the cylinder chamber.

Actuator - servo valve assembly

CNC System. The hydraulic press was controlled by the in-house developed Open Architecture CNC System [84, 76].

5.8.2   Dynamic Model of Hydraulic Actuator Module

An accurate dynamic model of the overall hydraulic system is rather difficult to obtain, owing to leakage in seals and connections, losses within flexible hoses, compressibility of the fluid and changes in its viscosity, and wear in the orifices and ports. In addition the flow-pressure expressions

Figure 5.24: Electro-hydraulic actuator assembly.

are nonlinear for a practical control algorithm design in a CNC system. However, even with some practical assumptions, a mathematical model of the hydraulic system is essential for the optimal design of hydraulic circuit and for selecting components. Such an analysis has been made in designing the hydraulic circuit for the press [84]. Only a simplified model of the hydraulic system is presented here in order to provide a basic foundation for a digital press brake controller design suitable for a CNC.

The model is based on the valve–actuator assembly shown in Figure 5.24. The second-order dynamics of the spool is much faster than the electronic torque motor, which has first-order dynamics. The transfer function between the spool valve displacement and supplied current to the torque motor is approximated as

$$\frac{x_v(s)}{I_a(s)} = \frac{K_i}{\tau_i s + 1}, \tag{5.129}$$

where K_i and τ_i are the gain and time constant of the torque motor, respectively. Assume that the secondary spool is at null flow position, where both ports connected to either sides of the cylinder are closed by the zero-overlap spool. The flow across the valve port is proportional to the area of opening and to the square root of the pressure drop. Since the area of valve opening is proportional to the spool displacement, the following expressions can be used to represent the flow through the ports for the extending and retracting motion of the piston [85]:

$$\left.\begin{aligned} q_a &= K_q x_v \sqrt{|P_s - P_a|} \operatorname{sign}(P_s - P_a), \\ q_b &= K_q x_v \sqrt{|P_b - P_t|} \operatorname{sign}(P_b - P_t), \end{aligned}\right\} \quad x_v \leq 0,$$

$$\left.\begin{aligned} q_a &= K_q x_v \sqrt{|P_t - P_a|} \operatorname{sign}(P_t - P_a), \\ q_b &= K_q x_v \sqrt{|P_b - P_s|} \operatorname{sign}(P_b - P_s), \end{aligned}\right\} \quad x_v > 0,$$

$$(5.130)$$

where P_s, P_t, P_a, and P_b are the supply, tank, and port (A,B) pressures, respectively.

The flow coefficients for ports A and B are assumed to be equal to K_q. The flows into the cylinder move the piston according to the conservation of mass on each chamber:

$$q_a = -A_p v_y + \frac{V_a}{\beta} \frac{dP_a}{dt} + K_{lp}(P_a - P_b),$$

$$q_b = -A_r v_y - \frac{V_b}{\beta} \frac{dP_b}{dt} + K_{lp}(P_a - P_b),$$

$$(5.131)$$

where β and K_{lp} are the bulk modulus of the fluid and the coefficient of leakage past the piston, respectively. In the particular single-rod piston system, the piston areas (A_p, A_r) on each side of the chamber are different. The term with a differential pressure and fluid bulk modulus represents the change in the chamber volume due to the compressibility of the fluid.

The fundamental flow equations for the servo valve and actuator assembly given above (Eqs. 5.130–5.131) have several nonlinearities besides the square root term. The expressions are linearized here for simplicity in deriving the transfer function of the overall hydraulic actuator system. The following assumptions are made: Both sides of the piston area are equal (i.e., symmetric actuator with $A_p = A_r = A$); the tank is open to the atmosphere and its relative pressure is zero, $P_t = 0$; the flows into and out of the spool valve or cylinder chambers are equal. Defining the load flow as $q_l = q_a = q_b$, and the load pressure as $P_l = P_a - P_b$, we can obtain the following relationships from Eq. (5.130):

$$K_q x_v \sqrt{|P_s - P_a|} \operatorname{sign}(P_s - P_a) = K_q x_v \sqrt{|P_b|} \operatorname{sign}(P_b),$$

which leads to

$$P_s = P_a + P_b, \quad P_l = P_a - P_b, \quad \rightarrow \quad P_b = \frac{P_s - P_l}{2}.$$

Substituting $P_t = 0$ and P_b into q_b (Eq. 5.130) yields

$$q_l = (K_q \sqrt{P_s/2}) x_v \sqrt{1 - P_l/P_s}.$$

$$(5.132)$$

The load flow is linearized by assuming small changes in the valve displacement (x_v) and load pressure (P_l). Linearizing q_l about the null flow position of the spool, we get

$$q_l = \left.\frac{\partial q_l}{\partial x_v}\right|_{P_{lo}, x_{v0}} x_v + \left.\frac{\partial q_l}{\partial P_l}\right|_{P_{lo}, x_{v0}} P_l$$

$$= K_{qo} x_v + K_{xo} P_l,$$

$$(5.133)$$

where the normalized valve flow (K_{qo}) and pressure gains (K_{xo}) are

$$K_{qo} = K_q \sqrt{\frac{P_s - P_{lo}}{2}}, \quad K_{xo} = \frac{-K_q x_{v0}}{2\sqrt{2(P_s - P_{lo})}}.$$

The flow equations in the cylinder (Eq. 5.131) can also be linearized. If we let the total volume in the cylinder be V_t and the volume displaced in the chambers be ΔV, the volumes in the chambers become $V_a = V_t/2 + \Delta V$ and $V_b = V_t/2 - \Delta V$. If we add the flows in each chamber (i.e., $q_a + q_b = 2q_l$), and assume a constant supply pressure P_s and equal piston areas ($A_p = A_r = A$), and consider that

$$\frac{dP_l}{dt} = \frac{dP_a}{dt} - \frac{dP_b}{dt}, \quad \frac{dP_s}{dt} = \frac{dP_a}{dt} + \frac{dP_b}{dt} = 0,$$

the load flow from equation (5.131) becomes

$$q_l = -Av_y + \frac{V_t}{4\beta}\frac{dP_l}{dt} + K_{lp}P_l. \tag{5.134}$$

The equation of motion for the equivalent mass (M_r) at the piston rod with negligible viscous friction is

$$\sum F_{\leftarrow}^{+} = M_r \frac{dv_y}{dt} = A(P_b - P_a) + F_l, \tag{5.135}$$

where the load $F_L = F_{ext} - F_c \, \text{sign}(v_y)$. Here F_{ext} and F_c are the sheet metal bending and Coulomb friction forces, respectively. The transfer function of piston velocity can be derived by combining the Laplace transforms of Eqs. (5.133), (5.134), and (5.135):

$$v_y(s) = \frac{K_x}{s^2 + 2\zeta_l \omega_l s + \omega_l^2} x_v + \frac{K_F(\tau_F s + 1)}{s^2 + 2\zeta_l \omega_l s + \omega_l^2} F_l, \tag{5.136}$$

where

$$\omega_l = 2A\sqrt{\frac{\beta}{V_t M_r}}, \quad \zeta_l = \frac{K_{lp} - K_{xo}}{A}\sqrt{\frac{\beta M_r}{V_t}},$$

$$K_x = \frac{-4\beta AK_{qo}}{V_t M_r}, \quad \tau_F = \frac{V_t}{4\beta(K_{lp} - K_{xo})}, \quad K_F = \frac{1}{\tau_F M_r}.$$

The open-loop transfer function indicates the influence of the external bending load and Coulomb friction on the ram velocity; the damping coefficient is proportional to the valve and piston leakage; and the stiffness of the actuator is proportional to the bulk modulus of the fluid and piston area while being inversely proportional to the stroke. The linearized transfer function is useful for analyzing the general behavior of the actuator for a given set of initial design parameters. However, the nonlinearities and the unknown values of fluid viscosity, friction, and leakage in various seals and ports require the on-line identification of the overall system transfer function, so that a well-tuned digital control algorithm can be embedded into the CNC computer.

5.8.3 Identification of Electro-Hydraulic Drive Dynamics for Computer Control

Although mathematical modeling is a useful tool for selecting hydraulic components, and determining the general order of the assembled system, the effects of valve deadband, hysteresis, stiction, leakage, and transport delay make the accurate modeling of the dynamics of hydraulic system for computer control difficult. A series of system identification tests are usually conducted to evaluate the discrete transfer function of the system for computer control.

Several characteristics of the hydraulic system must be considered in designing the identification experiments. Although the gain of the closed-center spool-valve is largest at the null-flow position, the deadband and hysteresis are also dominant. In addition, the amplitude of the input signal must be large enough so that the steady state flow gain identified is not biased by the valve-spool stiction. Because there is a significant Coulomb friction between the ram and guideways, the input excitation signal chosen must not reverse the motion and thus the static friction load. Furthermore, because we are using a single-rod asymmetrical actuator, the dynamics of the system in extending and retracting will not be the same. Finally, overall flow-pressure relationships are nonlinear owing to square root terms, and their linearization is acceptable only at small regions of actuator movements. Identification tests are conducted by considering the above mentioned characteristics of the hydraulic system. First, a series of step response tests are conducted in extending and retracting pistons separately. An approximate order and time delay of the system was estimated, which was later enhanced by PRBS (Pseudo Random Binary Sequence) excitation and least squares identification techniques. In both excitations, the minimum input signal amplitude was at least 3% of the maximum amplitude required to overcome the effects of nonlinear valve gain and spool friction. A dc offset is added to the input signal to prevent any reversal of piston velocity (and Coulomb friction load), which constrains the identification to either extending or retracting.

The input signals, and the corresponding actuator position, velocity, and pressure outputs, are collected synchronously with a multichannel digital scope. The digital data are later transferred to a computer and processed by MATLAB's Systems Identification Tool Box [1]. The digital position feedback, the encoder feedback to the CNC, is disabled, and input signals are sent to servovalve amplifiers as analog voltages via D/A converters, which have a gain of $K_{\mathrm{d}} = 20/2^9$ [V/count/s]. The velocities of both actuators are measured by magnetic field–based velocity transducers instrumented to the tie-rod slider mechanism. The data sampling frequency was fixed at 1 kHz during identification tests.

The open-loop discrete transfer function of the system was identified from the sampled values of PRBS inputs to the servovalve amplifier and outputs of the velocity transducer at 1 ms time intervals, which are equivalent to loop-closure periods. The system's open-loop discrete transfer function can be represented in the form of

$$G_o(z^{-1}) = \frac{v_y(z^{-1})}{u(z^{-1})} = \frac{z^{-d}B(z^{-1})}{A(z^{-1})}, \tag{5.137}$$

where d is the system's deadtime and

$$B(z^{-1}) = b_0 + b_1 z^{-1} + b_1 z^{-2} + \cdots + b_{n_b} z^{-n_b},$$
$$A(z^{-1}) = 1 + a_1 z^{-1} + a_2 z^{-2} + \cdots + a_{n_a} z^{-n_a}.$$

Alternatively, the system's response – the piston velocity (v_y) at time interval k – can be expressed in ARMA (Auto Regressive Moving Average) form as

$$v_y(k) = \{\phi(k)\}^T \{\theta\}, \tag{5.138}$$

where the measurement vector is

$$\{\phi(k)\}^T = [u(k{-}d)\, u(k{-}d{-}1) \cdots u(k{-}d{-}n_b), -v_y(k{-}1) \cdots -v_y(k{-}n_a)] \tag{5.139}$$

and the transfer function parameter vector is

$$\{\theta\} = [b_0\, b_1\, \cdots\, b_{n_b}\ a_1\, a_2\, \cdots\, a_{n_a}]^T. \tag{5.140}$$

The values of the transfer function, or the components of time-invariant parameter vector $\{\theta\}$, are estimated from N sets of measurements using an off-line standard least squares (LS) method, which is summarized in Appendix B. The MATLAB Identification Tool Box [1] has the LS identification routines used here. The input signal amplitude was chosen to be as 600 mV, or 3% of the maximum valve input in order to overcome valve stiction. In addition, this value is well within the range of valve input signals during the forming operation.

It was found that a first-order model represents the system dynamics at low frequency band, while a third-order model captures the dynamic mode contributed by the load and actuator at frequencies above 100 Hz. The delays in the system vary between 4 and 5 (ms) for each model. The left and right actuators have different dynamic characteristics, which reflect the machining and friction characteristics of each actuator–guideway assembly. In addition, each actuator exhibits different dynamics in extending and retracting motions owing to its asymmetrical piston. Although high-order models represent the system dynamics over a wide frequency range, a first-order model with five time delays is satisfactory for the extending motion of the actuators over the low frequency band where the forming operation takes place. The identified open-loop transfer function between the actuator velocity (v_y) and valve amplifier input (u) is expressed as

$$\frac{v_y}{u} = \frac{z^{-5}(b_0 + b_1 z^{-1})}{1 + a_1 z^{-1}}. \tag{5.141}$$

The excitation of high-order modes are avoided by applying smooth velocity changes using the trapezoidal velocity profile as shown in Figure 5.25.

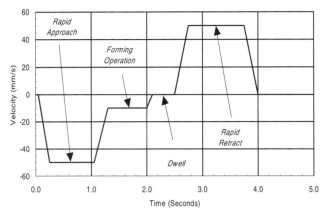

Velocity profile for a sheet metal bending cycle.

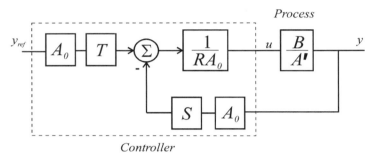

Block diagram of a pole-placement control system.

Figure 5.25: Press brake control system.

The reader is reminded of the following for the open-loop transfer function parameters of the press shown in Table 5.5:

- $a_1 = a'_1 + 1$, since a'_1 belongs to position (see next section).
- The values b_0 and b_1 should be multiplied by the digital to analog converter gain of the CNC to obtain the correct transfer function units between position input and output in *counts*. A nine-bit two-parallel pulse width modulated circuit with ± 10 V is used as an analog to digital converter. The gain of the digital to analog converter circuit is $K_d = 20/2^9 = 0.0391$ [V/(count/s)].

5.8.4 Digital Position Control System Design

We need to design a digital control algorithm with a minimum computation time and high control loop closure frequency while providing a low positioning error. The performance of the positioning system during the bend and dwell operations determines the accuracy of the press. In these cases, the actuators are either extending or holding their position against a forming load. Therefore the first-order models that represent the velocity dynamics during actuator extensions are used for the digital control of the press ram position. The position

TABLE 5.5. Pole-Placement Axis Control Law Parameters for the Hydraulic Press. Note that the D/A Converter Gain is $k_d = 20/2^9$ [V/counts/s]

$[A_m] = 1 + a_{m1}z^{-1} + a_{m2}z^{-2}$	$1 - 1.429\,z^{-1} + 0.4724\,z^{-2}$ Desired Model
Left Axis – y_1	**Pole-Placement Design Parameters**
$[B] = b_0 + b_1 z^{-1}$	$(1573 - 702z^{-1})$ [count/V] $\cdot k_d$
$[A] = 1 + a'_1 z^{-1} + a'_2 z^{-2}$	$1 - 1.871z^{-1} + 0.871z^{-2}$
$[T] = t_0$	4.8837×10^{-4}
$[S] = s_0 + s_1 z^{-1}$	$0.00444 - 0.00395z^{-1}$
$[R] = r_0 + r_1 z^{-1} + \cdots + r_5 z^{-5}$	[1 0.442 0.4284 0.4165 0.4062 0.1244]
Right Axis – y_2	**Pole-Placement Design Parameters**
$[B] = b_0 + b_1 z^{-1}$	$(982 - 505z^{-1})$ [count/V] $\cdot k_d$
$[A] = 1 + a'_1 z^{-1} + a'_2 z^{-2}$	$1 - 1.861z^{-1} + 0.861z^{-2}$
$[T] = t_0$	7.4717×10^{-4}
$[S] = s_0 + s_1 z^{-1}$	$0.00645 - 0.0057z^{-1}$
$[R] = r_0 + r_1 z^{-1} + \cdots + r_5 z^{-5}$	[1 0.432 0.4154 0.401 0.3887 0.1306]

of the ram is measured with a linear encoder attached to the ram-tie-rod assembly. The linear encoder and position counter in the CNC act as an integrator $(1/(1-z^{-1}))$ and provide the actual position of the actuator end point at discrete control intervals $T_s = 1$ ms. The resulting open-loop transfer function between the amplifier input (u [V]) and actuator position (y) becomes

$$\frac{B(z^{-1})}{A'(z^{-1})} = \frac{y}{u} = \frac{v_y}{u}\left(\frac{1}{1-z^{-1}}\right) = \frac{z^{-5}(b_0 + b_1 z^{-1})}{1 + a'_1 z^{-1} + a'_2 z^{-2}}, \tag{5.142}$$

where $a'_1 = a_1 - 1$ and $a'_2 = -a_1$. Here, (u) is in Volts and y unit is in counts, where 1 count = 0.005 mm. The open position loop transfer function can be expressed in an alternate form using a forward shift operator (z) for convenience:

$$\frac{y(k)}{u(k)} = \frac{B(z^{-1})}{A'(z^{-1})} = \frac{b_0 z + b_1}{z^4(z^2 + a'_1 z + a'_2)} \quad \left[\frac{\text{counts}}{V}\right]. \tag{5.143}$$

The above transfer function has a first-order numerator (i.e., $\deg(B(z)) = 1$) and a sixth-order denominator (i.e., $\deg(A(z)) = 4 + 2 = 6$). A pole-placement control scheme presented by Astrom [86] allows performance-based design, where the performance criteria are 1) to have a high gain to reject bending and friction load disturbances, 2) to compensate delays in the open-loop transfer function, and 3) to have low steady state positioning errors. The block diagram of the pole-placement controller is shown in Figure 5.25, where the input is the command and output is the actual position of the actuator, respectively. B/A is the open-loop transfer function of the system given in Eq. (5.141). The digital controller consists of a feed forward filter $T(z)$, a feed back filter $S(z)$, controller poles $R(z)$, and an observer $A_0(z)$. The transfer function of the closed-loop control system can be derived from the block diagram (Fig. 5.25):

$$\frac{y(k)}{y_{\text{ref}}(k)} = \frac{(BT)/(A_0 A' R)}{1 + (B A_0 S)/(R A_0 A')} = \frac{BT}{A'R + BS}. \tag{5.144}$$

The control polynomials R, S, T, and A_0 are designed in such a way that the closed-loop response of the overall system behaves like a desired model transfer function:

$$\frac{y(k)}{y_{\text{ref}}(k)} = \frac{BT}{A'R + BS} = \frac{A_0 B_m}{A_0 A_m},$$ (5.145)

where polynomials A_m and B_m contain the desired poles and zeros of the system. The desired system response was chosen to have second-order dynamics, with a settling time of $t_s \approx 4/(\xi \omega_n)$ for underdamped and $t_s \approx -\ln 0.05/(\omega_n(\xi - \sqrt{\xi^2 - 1}))$ for overdamped systems. The desired characteristic equation is selected as

$$A_{m0} = s^2 + 2\xi \omega_n s + \omega_n^2 = s^2 + 750s + 62500$$

with a natural frequency of 250 rad/s. The roots of the overdamped ($\xi > 1$) characteristic equation are

$$s_{i,ii} = \omega_n(-\xi \pm \sqrt{\xi^2 - 1}) = (-95.4915, -654.5085).$$

The corresponding discrete time domain roots for a $T = 1$ ms sampling interval become

$$z_i = e^{s_i T} = 0.9089, \quad z_{ii} = e^{s_{ii} T} = 0.5197,$$

with the desired discrete transfer function of

$$A_{m0} = (z - z_i)(z - z_{ii}) = z^2 - 1.429z + 0.4724 = z^2 + a_{m1}z + a_{m2}.$$

The resulting desired model transfer function is selected to have the same numerator (i.e., zeros) and the same order of the denominator of the open-loop process dynamics:

$$\frac{B_m}{A_m} = \frac{y(k)}{y_{\text{ref}}(k)} = \frac{b_{m0}(b_0 z + b_1)}{z^4(z^2 + a_{m1}z + a_{m2})} \left[\frac{\text{count}}{\text{count}}\right],$$ (5.146)

where y_{ref} is the commanded reference position and b_{m0} is a scale factor to ensure a unity gain in the overall closed-loop system.

$$\left|\frac{B_m(z)}{A_m(z)}\right|_{z=1} = 1, \quad \rightarrow \quad b_{m0} = \frac{1 + a_{m1} + a_{m2}}{b_0 + b_1}.$$

Based on the design criterion of causality rules (i.e., the present input must not depend on the future outputs of the system) proven by Astrom and Wittenmark [86], the control polynomials must have the following orders:

$$\deg(A_0(z)) \geq 2 \deg(A'(z)) - \deg(A_m(z)) - 1 = 5, \quad \text{let } \deg(A_0(z)) = 5,$$
$$\deg(R(z)) = \deg(A_0(z)) + \deg(A_m(z)) - \deg(A'(z)) = 5 + 6 - 5 = 5,$$
$$\deg(S(z)) \leq \deg(R(z)), \quad \text{let } \deg(S(z)) = 5,$$
$$\deg(T(z)) = \deg(A_0(z)) = 5.$$

The corresponding control polynomials are as follows

$$A_0 = z^5,$$
$$R(z) = z^5 + r_1 z^4 + r_2 z^3 + r_3 z^2 + r_4 z + r_5, \tag{5.147}$$
$$S(z) = s_0 z^5 + s_1 z^4 + s_2 z^3 + s_3 z^2 + s_4 z + s_5.$$

By considering the equivalence of the actual closed-loop and desired transfer function models (Eq. 5.145), we can identify the parameters of the control polynomials from the solution of following Diophantine equations [86]:

$$B(z)T(z) \equiv A_0(z)B_m(z),$$
$$A'(z)R(z) + B(z)S(z) \equiv A_0(z)A_m(z). \tag{5.148}$$

By considering the numerator terms, we have

$$(b_0 z + b_1)T(z) \equiv z^5 b_{m0}(b_0 z + b_1) \rightarrow T(z) = b_{m0} z^5. \tag{5.149}$$

When $A'(z)R(z) + B(z)S(z) \equiv A_0(z)A_m(z)$ is considered, we get

$$
\begin{aligned}
z^4 (z^2 &+ a'_1 z + a_2)(z^5 + r_1 z^4 + r_2 z^3 + r_3 z^2 + r_4 z + r_5) \\
&+ (b_0 z + b_1)(s_0 z^5 + s_1 z^4 + s_2 z^3 + s_3 z^2 + s_4 z + s_5) \\
&\equiv z^5 z^4 (z^2 + a_{m1} z + a_{m2}).
\end{aligned}
$$

The resulting expressions for the parameters are

$$r_1 = a_{m1} - a'_1, \quad r_2 = a_{m2} - a'_2 - a'_1 r_1,$$
$$2 < j < (d = 5) \rightarrow r_j = -(a'_1 r_{j-1} + a'_2 r_{j-2}), \quad j = 3, 4,$$
$$r_d = r_5 = \left[\left(a'_1 \frac{b_1}{b_0} - a'_2 \right) r_{d-1} + a'_2 \frac{b_1}{b_0} r_{d-2} \right] \Big/ \left[a'_1 - a'_2 \frac{b_0}{b_1} - \frac{b_1}{b_0} \right],$$
$$s_0 = -\frac{r_d + a'_1 r_{d-1} + a'_2 r_{d-2}}{b_0}, \quad s_1 = -\frac{a'_2 r_d}{b_1}, \quad s_j = 0, \quad j = 2, \ldots, d,$$

where delay is $d = 5$. Since the parameters of the open-loop transfer function are assumed to be time invariant, the Diophantine equations are solved once during the design. The computed numerical values used for the right and left actuator position control are presented in Table 5.5.

From the block diagram of the control system, the position command input is generated according to the following expression:

$$R(z)u(k) = T(z)y_{\text{ref}}(k) - S(z)y(k). \tag{5.150}$$

Substituting the control polynomial parameters (R, T, S) leads to the following control law being executed at each loop closure time T_s:

$$u(k) = t_0 y_{\text{ref}}(k) - \sum_{j=0}^{1} s_j y(k-j) - \sum_{j=1}^{5} r_j u(k-j). \tag{5.151}$$

A series of position jumps are commanded to both actuators to test the step response behavior of the controlled press (see Fig. 5.26). It is observed that both actuators have approximately matched dynamic response, which is essential in

Step response of right and left actuators

Response of actuators to pulse inputs with variable amplitudes

Figure 5.26: The response of the right and left axis positioning systems to step changes in the position command.

coordinated precision bending, and the system has a total rise time of about 12 ms. Although the same control law was used in both extending (negative direction) and retracting motions, the actuators do not exhibit overshoots, which is a basic requirement for a machine tool control system. To test the influence of deadband and hysteresis, which are caused by the valve-spool stiction and

Coulomb friction in the guides, a series of pulse position commands with an increasing amplitude were demanded from the CNC system. The actuator positions were monitored at 2 ms intervals and are shown in Figure 5.26. The results indicate that the deadband of the controlled motion is within 1 encoder unit (0.005 mm), which is quite satisfactory.

The computer control strategy presented here is applicable to many other motion or process control applications found in practice.

5.9 PROBLEMS

1. CNC Design and Analysis Project: A simplified block diagram of a CNC milling machine's feed drive system is shown in Figure 5.27. The parameters of the system are given below:

$H_g = 0.0913$ V/(rad/s); Tachogenerator constant.
$K_d = 0.0781$ V/counts; D/A converter gain.
$K_e = 636.6$ counts/rad; Encoder gain.
$K_p = ?$; Digital filter gain.
$a = ?$; Digital filter's zero.
$b = ?$; Digital filter's pole.
$T = 0.001$ s ; Sampling period.

(a) The transfer function of the velocity loop $(G_v(s))$ is not known. Measure the transient response of the velocity loop by applying a

Figure 5.27: Block diagram and command position history of a feed drive control system.

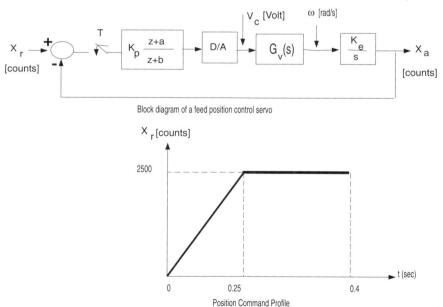

Block diagram of a feed position control servo

Position Command Profile

rectangular waveform with $V_c = 1\,\text{V}$ amplitude to the servoamplifier. The actual velocity of the feed drive is measured from the tachogenerator output, which has a gain of H_g [V/(rad/s)]. Identify the
transfer function of the velocity loop $(G_v(s) = \omega(s)/V_c(s) =?)$ by assuming that it can be approximated as a first-order system. Simulate the step response of the velocity loop in the continuous time domain using Laplace transforms, and compare the response with the
experimental measurements by plotting them on the same graph.
(Note: If experimental setup is not available, assume $G_v(s) = 11/$
$(0.008s + 1)$.)

(b) Measure the frequency response of the velocity loop. Plot the magnitude ratio and phase difference up to 100 Hz. Determine the bandwidth
of the velocity loop.

(c) Design the digital filter of the position control system in such a way
that the overshoot is negligibly small (i.e., $M_p = 1\%$) and the rise time
is about 20 ms. You can design the control law either in the s or the z
domain.

(d) Assuming that the CNC computer is sending velocity signals (V_c) at
$T = 1$ ms intervals to the servoamplifier via a D/A converter, identify
the transfer function of the entire position loop in the z or the s domain. Simulate the response of the position loop for the input shown
in Figure 5.27 using the discrete transfer function of the system. Identify the steady state error theoretically and compare it with the error
indicated by the simulation.

(e) Derive the state-space response of the position loop. Simulate the position and velocity responses of the position control system for the input
profile given in Figure 5.27. Compare the position response profile
with the transfer function simulation results obtained in the previous
question.

(f) Design a real time interpolation algorithm for a combination of linear and circular tool paths. Simulate the given tool paths on a PC. If
possible, try the algorithms on a real machine tool in a laboratory.

(g) Combine the real time interpolation and the time domain feed drive
servo simulation algorithms. Analyze the contouring errors on a circular tool path by increasing the feed velocities.

2. A High Speed Linear Motor Driven XY Table Design: A high speed, two-
axis XY table is driven by linear (i.e., flat bed) dc motors and linear guides.
The assembly drawing and the block diagram of the continuos velocity
control loop are provided in Figure 5.28. The actual position (x_a [m]) of
the head is measured with a linear encoder. The position loop is closed in
the digital computer, and the position error is passed from a digital filter
$D(z) = K_p \frac{z+a}{z+b}$ at T discrete time intervals. The digitally filtered error is
sent to a linear motor amplifier via a 12-bit D/A converter with a voltage
range of ± 10 V. The x axis of the table has the following parameters:

$K_i = 9{,}700$ A/A, $K_p = 31.8$ V/v, $K_a = 0.25$ V/A, $K_b = 13.78$ V/(m/s), $L_a =$ 0.75 mH, $R_a = 2.8\,\Omega$, $T_g = 0.068$ V/V, $H_g = (1$ V)/(0.03 m/s), $H_i = 5.246$, $K_t = 15.568$ V/V, $K_e = 1$ μm/count. The saturation limits of the PWM voltage amplifier are ±159 V, and the control interval is $T = 0.001$ s.

(a) Derive the transfer function of the velocity loop $(G_v(s))$.

(b) The inner velocity loop has a high bandwidth, thus it can be approximated as a gain $E = 4$ A/V as shown in Figure 5.28. The frequency response of the velocity loop is measured using the experimental setup shown. The magnitude of the system at frequencies $\omega = 0$, 1000 rad/s were measured to be $|G_v(0)| = 0.1076$ m/s/V and $|G_v(1000)| = 0.0139$ m/s/V, respectively. Estimate the table mass $(M_a$ (kg)) and reference signal gain (S_g). What are the gain and the time constant of the velocity loop?

(c) Design the digital filter $(D(z))$ in such a way that the overall closed-position loop behaves like a second-order *overdamped* system with two negative real poles, $s_1 = -100$ rad/s, $s_2 = -150$ rad/s. $(K_p, a, b =?)$

(d) Find the steady state error of the position control loop for a steady state feed velocity of $f = 0.1$ m/s.

(e) Remove the velocity feedback from the system $(T_g = H_g = 0)$, and derive the closed-loop transfer function of the complete position loop (in the s or the z domain). Consider the inner loop in derivations.

(f) Simulate current, velocity, position, and velocity command (V_c) for a unit step and ramp inputs with a constant Coulomb friction of 10 N. Control software tools can be used for the simulation.

3. CNC Hydraulic System Design Project: You are assigned to design a digital control system for a hydraulic actuator driving a lumber bandsaw head. The physical control system consists of an amplifier driving a servovalve–piston/cylinder actuator carrying the bandsaw head. The actual position $(x_a$ [μm]) of the head is measured with a linear encoder circuit, which has a gain of K_e [μm/mm], and the linear velocity of the head motion is monitored in [mm/s] by a velocity transducer. The position loop is closed in the digital computer, and the pole placement control law explained in Section 5.8.4 is used for the position control of the actuators. The resulting control variable is sent to a servovalve amplifier via a D/A converter, which has a gain of K_d. The input voltage to the velocity loop $(V_c$ [V]) is received from the D/A converter, and the output velocity is measured from a velocity transducer in u [mm/s]. Thevelocity loop is found to behave like a first-order continuous system with a gain of K_v [(mm/s)/V] and time constant of τ_v [s].

(a) Draw the block diagram of the complete control system.

(b) Derive the state-space equations for the velocity u [mm/s] and head position x_a [μm]. Neglect the disturbance load.

(c) The gain and the time constant of the velocity loop were estimated from a simple frequency response test by sending V_c [V] $= \sin 100t$ to the servovalve amplifier and measuring the resulting sinusoidal

Experimental set-up for measurements

Simplified block diagram of one drive without encoder and CNC computer. The inner loop is E.

High speed XY table with linear drives and motors.

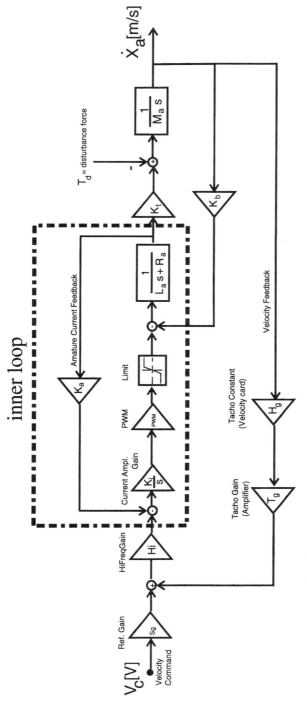

Figure 5.28: A high speed linear XY table system. Parts: 1,2 = dc linear motors for x and y axis; 3,4 = linear encoders for x, y drives; 5 = top positioning plate; 6 = spring loaded bumper; 7,8 = linear support rails for x, y drives.

velocity response. The measured velocity was observed to have a maximum amplitude of $u = 70.71$ mm/s and a time delay of $T_d = 0.0078$ s with respect to the sinusoidal input V_c. What are the gain and the time constant of the velocity loop?

(d) The control interval is $T = 0.001$ s, the encoder circuit has a resolution of 1,000 μm/mm (i.e., 1 count $= 1 \mu$m), and the D/A circuit has a 10-bit resolution with a voltage range of ± 5 V. Design the pole-placement control law in such a way that the overall closed position loop behaves like a second-order underdamped system with a natural frequency of $\omega_n = 250$ rad/s and a damping ratio of $\zeta = 1.5$.

(e) Express the states (u [mm/s], x_a [μm]) as difference equations (from item b) with numerical values. Assume the input to be series of $x_r(k)$.

4. Sample Interpolation Problems: Assume a machine tool system with equal acceleration and deceleration rates of $A = D = 20,000$ count/s^2, and a minimum interpolation time of $T_{min} = 0.002$ s. The position and feed units are in *counts* and *counts/s*, respectively.

(a) Illustrate the working principle of a linear interpolator for an NC block:

N010 G91 G01 X5 Y $-$ 3 F6000

Calculate the interpolation parameters and show the resulting tool path on a graph and in a table. (Note: All units are in *counts*, and feed is in *counts/s*.)

(b) Illustrate the working principle of a circular interpolator for an NC block

N010 G91 G02 X60 Y $-$ 30 I $-$ 60 J $-$ 30 F10000

Calculate the interpolation parameters and show the resulting tool path on a graph and in a table. (Note: All units are in *counts*, and feed is in *counts/s*.)

CHAPTER SIX

SENSOR-ASSISTED MACHINING

6.1 INTRODUCTION

The first step in automating machining systems was the introduction of CNC machine tools. The primary function of CNC is to automatically execute a sequence of multiaxis motions according to a part geometry. However, safe, optimal, and accurate machining processes are generally planned by manufacturing engineers based on their experience and understanding of the process. It is rather difficult to predict vibration, tool wear and breakage, thermal deformation of the machine tools, and similar process based on events using off-line theoretical models. In addition to engineering the process plans before actual machining, the machine tools are instrumented with vibration, temperature, displacement, force, vision, and laser sensors in order to improve the productivity and reliability of the cutting operations on-line. The sensors must have reliable frequency bandwidth, have good signal to noise ratio, and provide signals with reliable correlation to the state of the process. They must also be practical for installation on machine tools. The measured sensor signals are processed by real time monitoring and control algorithms, and the corrective actions are taken by the CNC accordingly. The corrective actions may be manipulation of spindle speed, feed, tool offsets, compensation of machine tool positions, feed stop, and tool change depending on the process monitoring and control application. Such a sensor-assisted cutting is called *intelligent machining* in the literature [87, 88]. The architecture of CNC must be organized in such a way that it allows real time manipulation of the machine tool's operating conditions. In other words, CNC must be *open* to allow integration of user-developed real time application programs. The following *Intelligent Machining Module* has been developed at the author's laboratory [76, 81] for modular integration of cutting process monitoring and control tasks to CNC machine tools.

6.2 INTELLIGENT MACHINING MODULE

An Intelligent Machining Module (IMM) has been designed to run on existing commercial CNC systems that allow limited manipulation of cutting conditions by the end users. The IMM runs on a Digital Signal Processing (DSP) board with analog sensor signal processing capability. Various intelligent machining tasks, such as adaptive control, tool condition monitoring, and process control, can simultaneously run on the system. The user can reconfigure the system by using script commands from the supplied signal processing and data collection library. Each function is called a Plug In Module (PIM), and the IMM has a mechanism of integrating new, user-developed PIMs to the script command library. The IMM is configured to communicate with commercial semi-open CNCs through the PC–CNC communication links and software. Several IMM systems have been installed in industry. Some examples include adaptive control, chatter detection, and tool failure monitoring applications on a five-axis machining center controlled by a FANUC^{TM} CNC with a PC interface. The IMM sends feed and spindle speed change, machining halt, tool change, tool offset, and other NC commands accepted by the CNC controllers.

6.2.1 Hardware Architecture

Real time IMM algorithms can run on any DSP board that has a shared memory with the host PC and analog channels to collect sensor signals. Analog outputs of the sensors, such as cutting force, vibration, temperature, and pressure, sensors, can be connected to the DSP board's data acquisition module (see Fig. 6.1). The DSP board is controlled by PC software running under Windows NT operating system. Some commercial CNC systems either allow high speed communication with an external PC or they have an internal PC. Machine tool commands, such as *feed change, spindle speed change, tool offset, tool change*, and similar commands can be sent from C programs running on the PC to the executive of the commercial CNC. Such CNC systems, which allow limited interface with user applications, are called "semi-open CNC systems" in this book. The IMM has been designed to run on semi-open CNC systems.

Figure 6.1: Sensor-assisted intelligent machining system.

6.2.2 Software Architecture

The IMM is a combination of a PC system manager running on the host PC and a DSP system manager running on each DSP board in the PC/ISA bus. The DSP system manager is configured to act as server, while the PC manager acts as a client. All real time sensor data collection and signal processing activities are handled in the DSP. The PC system manager initializes the user-requested functions, sets up signal processing networks, and exchanges information among the DSP, user, and CNC. In order to eliminate any hardware dependency, a generic DSP interface was developed as a software module on the host PC. Each DSP driver supplied by the DSP manufacturer communicates with the rest of the software via the generic DSP interface, which is hardware independent. The generic DSP interface provides high level services such as command buffers, immediate data transfer protocol, and circular and double buffers for data transfer. The IMM also features a generic I/O interface in the DSP software, for high level access to analog I/O, interrupts, and timers, and it has a low level I/O driver for DSP-dependent operations. Different I/O and DSP boards can easily be supported by simply changing hardware-dependent drivers in the PC and the DSP software without modifying any user-developed functions.

Signal Processing Networks (SPNs). The SPN is a collection of application-specific tasks, which are called plug in modules (PIMs) in the IMM as shown in Figure 6.2.a. Each SPN runs on a single DSP board. The PIMs in an SPN are executed sequentially, which eliminates the need for explicit synchronization, thereby enabling one to use a very fast communication scheme without endangering the integrity of the transferred data. An SPN can have multiple levels running at different frequencies, so that data collection and processing can be done at one frequency, and additional processing and decision making can be done at another frequency. The PIMs and the links between the PIMs can be configured at runtime using a high level script language [76].

Plug in Module (PIM). The PIM interface allows application-specific tasks to be implemented at an abstract level without knowing the hardware details of the underlying system. Functions such as filters, control algorithms, transfer function identifiers, and fast Fourier transforms can be implemented as PIMs so that they can be integrated with the rest of the system. The inputs and outputs of a PIM are shown in Figure 6.2.b.

6.2.3 Intelligent Machining Application

The following script commands are set by a user for adaptive cutting force control and tool breakage detection in a five-axis milling application. Cutting force sensors, with a calibration factor of 1,000 N/V, are connected to analog channels 0, 1, and 2 of the DSP board. Cutting forces are sampled at 2,000 Hz, they are low pass filtered at 300 Hz, and the peak and average resultant forces are identified. There are four flutes on the end mill, and the spindle speed is 600 rev/min (10 Hz). In the second level, which corresponds to tooth passing

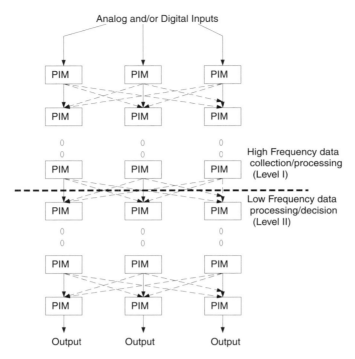

a) Signal Processing Network (SPN) structure.

b) Plug In Module (PIM) structure.

Figure 6.2: Real time information flow structure in intelligent machining module.

frequency (40 Hz), the peak cutting force is used by the adaptive force control algorithm and the average force is used by the tool breakage detection function. The adaptive control PIM calculates the required feed to keep the force constant, and the tool failure algorithm sends a feed hold signal if chipping is detected. Both algorithms pass the feed and feed hold commands to the CNC via a PC handler.

The system is set up via the following script commands:

Sensors are connected to analog channels 0,1, and 2 of DSP board.
INPUT SCALE 0, 1, 2; 1000, 1000, 1000 // Calibration: 200 N/V

```
BEGIN 0,1,2; 2000 // Sample input channels 0, 1, and 2 at 2,000 Hz.
LOWPASS FILTER(0; 16; 300) // Apply low pass filter of length 16.
LOWPASS FILTER(1; 16; 300) // with a cutting frequency of 300 Hz
LOWPASS FILTER(2; 16; 300) // to channels 0, 1, and 2.
PEAK DETECT(0,1,2 -> 5 ; ; )  // Write peak and average of force vector
AVERAGE DETECT(0,1,2 -> 7; ; ) // available on channels 0, 1, and 2 to
                              // software channels 5 and 7.
SECOND LEVEL Timer 40      // Start 2nd algorithm level to run at 40 Hz
// Run adaptive control using peak force available on channel 5.
   ADAPTIVE CONTROL ( 5, 6 ; ; 200, 200, 4, 0.2, 0.05 )
// Run tool breakage detection using average force on channel 7.
   TOOL BREAKAGE DETECTION ( 7, 8 ; ; )
IMMEDIATE OUTPUT TO PC ( 6, 8 ; ; ) // Send the calculated feed and
                              // feed hold commands to the PC.
DOUBLE BUFFERED OUTPUT TO PC( 5, 6 ; ; ) // Store peak force and
                              // identified feed in a buffer.
END                           // End second loop.
// Send feed override value to semi-open CNC.
IMMEDIATE HANDLER = Write feed override and emergency stop (0.05)
// Save all data in a disk file.
DOUBLE HANDLER = Write to disk (c:results.dat)
```

The graphical user interface of the IMM system records any signal or parameter indicated within the script command file. A sample window of commanded and measured cutting forces is shown in Figure 6.3. The fundamental philosophy behind the IMM system is that it allows integration of new, user-developed functions (i.e., PIMs) to the script library, and a new application can rapidly be configured by calling script commands in the desired sequence. The mathematical details of sample intelligent machining tasks – *adaptive cutting force control, tool breakage detection*, and *chatter avoidance* – are provided in the following sections.

Figure 6.3: Peak cutting force controlled by an adaptive control PIM.

6.3 ADAPTIVE CONTROL OF PEAK FORCES IN MILLING

6.3.1 Introduction

There are several physical constrains in the machining of metals. Some of these constrains are machine tool dependent, such as maximum torque and power available from the spindle drives. Other constraints are tool and workpiece dependent. Overloading of a carbide insert must be avoided by keeping the maximum chip thickness removed at a level that would not increase the principal tensile stress in the cutting wedge beyond the ultimate tensile strength of the tool material. Maximum static deflections left on the finish surface must be within the tolerance of the workpiece. Maximum resultant cutting force on a slender end mill must be kept safely below a limit value to prevent fracture of the shank. Here we present two different adaptive control strategies, which constrain the maximum resultant force at a safe level by adaptively manipulating the chip load or the feed rate.

A general block diagram of a typical adaptive control system is shown in Figure 6.4. The input to the system is the reference or desired level of the maximum cutting force. The actual cutting forces are measured via sensors mounted on the table or on the spindle and are collected at 3 to 5 degrees of angular rotation intervals to capture the peak resultant force. The peak force at each tooth or spindle period is evaluated and passed to the adaptive control law. When there are multiple teeth on the cutter, it is recommended to use peak forces per spindle revolution. Otherwise run-outs on the cutter would give fluctuating peak forces at each tooth period. The adaptive control would send similarly fluctuating feed rates to the machine tool drives, which have a low bandwidth. As a result, the adaptive control system would produce an oscillatory response, which is not desirable. Hence the peak force ($F_p(k)$) is evaluated at each spindle period (k) and subtracted from the set or desired force level ($F_r(k)$). The adaptive control algorithm determines a new feed command to minimize the force error. The feed command ($f_c(k)$) is sent to the CNC unit, which has its own digital position control law executed at comparatively small time intervals (e.g., 0.100 ms). The CNC unit sends voltage to the feed drive motors, which move the table at an actual feed velocity of f_a [mm/s]. Since machine tool drive control servos are tuned to be overdamped with zero overshoot, they can be approximated

Figure 6.4: Block diagram of a general adaptive control system in machining.

to have first-order dynamics with an average time constant of 0.1 ms. Note that this value may depend on the type of the machine tool and can be quite small for high speed machine tool drives. The cutting process feels the change in the chip load or the peak cutting force ($F_p(k)$) at least after one tooth period. The chatter vibration–free cutting process can be approximated as a first-order system with a time constant equal to one or more tooth periods but less than the spindle period. As the width or depth cut changes on the workpiece, the peak cutting force varies along the tool path. Hence, the first-order dynamics of the process has time-varying parameters. The combined CNC, machine tool feed drive, and cutting process can have an approximate second-order dynamics ($G_c(s)$). The peak force is evaluated from measurements and sent to an on-line *process identification algorithm*, which estimates the digital parameters of the combined machine tool, CNC, and time-varying cutting process as the cutter travels through holes, slots, and other features of the machined part. The time-varying coefficients of the process are estimated from the feed command input and peak force output of the process. Estimated parameters of the machine tool and cutting process are used to update the parameters of the adaptive control law at each control interval. Because the control law parameters are adjusted according to the changes in the cutting process parameters at each sampling interval, the control system is *adaptive* to the varying workpiece geometry. An adaptive control law, running at the control computer, computes new feed-rate command value at each sampling interval. The feed command moves the machine tool table according to the servo dynamics of the feed drive. The actual feed velocity of the table changes the chip thickness and therefore the cutting force produced during machining. The adaptive control loop ensures that the actual force felt by the milling cutter is always equal to the reference constant force value which is safely below the breakage limit. In the following sections we present step-by-step the designs of a pole-placement controller and a generalized predictive adaptive controller.

6.3.2 Discrete Transfer Function of the Milling Process System

As explained above, the machine tool control and drive system can be approximated by a first-order continuous system [89]:

$$G_m(s) = \frac{f_a(s)}{f_c(s)} = \frac{1}{\tau_m s + 1}, \tag{6.1}$$

where the f_a and f_c are the actual output and command input values of the feed speed in [mm/s]. The feed or chip load per revolution can be found by $h[\text{mm/rev}] = f_c/(Nn)$, where N is the number of teeth on the milling cutter and n [rev/s] is the spindle speed. A turning process can be modeled by assigning $N = 1$. The cutting force does not change instantly with the feed, and the process can be approximated to have first-order dynamics:

$$G_p(s) = \frac{F_p(s)}{f_a(s)} = \frac{K_c a b(\phi_{st}, \phi_{ex}, N)}{Nn} \frac{1}{\tau_c s + 1}, \tag{6.2}$$

where K_c [N/mm^2], a [mm], and $b(\phi_{st}, \phi_{ex}, N)$ are the cutting constant, the depth of cut, and immersion function, respectively. The immersion function ($b(\phi_{st}, \phi_{ex}, N)$) does not have any units and may change between 0 and $\sim N$ depending on immersion angle and the number of teeth in cut. Both the axial depth of cut (a) and the radial depth of cut (b) may change along the tool path depending on the workpiece geometry; hence a and b are time varying. Note that when chatter vibrations are present, the process is very unstable producing large-amplitude, oscillating cutting forces. The process has complex, high-order nonlinear dynamics and can not be controlled by manipulating the feed with an adaptive control system. Chatter control must be treated separately, and the adaptive control must be deactivated when chatter vibrations are present.

Although the machine tool dynamics can be considered as time invariant, it is more convenient to treat a combined machine tool and cutting process as time varying for practical application of adaptive control on multiaxis machine tools. The combined transfer function of the system is

$$G_c(s) = \frac{1}{(\tau_m s + 1)} \frac{K_c a b}{N n (\tau_c s + 1)} = \frac{K_p}{(\tau_m s + 1)(\tau_c s + 1)}, \tag{6.3}$$

where the process gain is K_p[N/mm/s] $= K_c a b / (N n)$. In reality, τ_c may change owing to static deflections of the tool–workpiece structure, which in turn affect the chip load. However, the above approximation is valid for practical simulation studies of the adaptive control algorithms. Since the machining process is controlled at spindle period T, the zero-order hold equivalent of $G_c(s)$ is considered:

$$G_c(z) = \frac{F_p(k)}{f_c(k)} = (1 - z^{-1}) \mathcal{Z} \frac{G_c(s)}{s} = \frac{b_0 z + b_1}{z^2 + a_1 z + a_2}, \tag{6.4}$$

where k is the spindle rotation counter, z is the forward shift operator, and

$$b_0 = K_p \frac{\tau_m (1 - e^{-T/\tau_m}) - \tau_c (1 - e^{-T/\tau_c})}{\tau_m - \tau_c},$$

$$b_1 = K_p \frac{\tau_c (1 - e^{-T/\tau_c}) e^{-T/\tau_m} - \tau_m (1 - e^{-T/\tau_m}) e^{-T/\tau_c}}{\tau_m - \tau_c},$$

$$a_1 = -(e^{-T/\tau_m} + e^{-T/\tau_c}),$$

$$a_2 = e^{-T(1/\tau_c + 1/\tau_m)}.$$

The discrete process parameters (a_1, a_2, b_0, b_1) depend on the workpiece geometry, and they may change during machining. For example, during the machining of engine blocks or aircraft wings, both the width of cut (i.e., immersion) and the axial depth of cut may change according to the workpiece and selected tool path geometry.

In order to design an adaptive control system for regulating the maximum cutting forces, the unknown time-varying parameters (a_1, a_2, b_0, b_1) must be estimated at each spindle period. A recursive least square (RLS) algorithm is used to estimate the time-varying parameters [90]. Henceforth, $\hat{a}_1, \hat{a}_2, \hat{b}_0$, and $\hat{b}_1$ will be used for estimates of the cutting process parameters. The following RLS algorithm is executed in a sequential manner at each spindle period (k):

$$\hat{\theta}(k) = \hat{\theta}(k-1) + K(k)[F_p(k) - \varphi(k)^T\hat{\theta}(k-1)], \tag{6.5}$$

$$K(k) = \frac{P(k-1)\varphi(k)}{\lambda + \varphi(k)^T P(k-1)\varphi(k)}, \tag{6.6}$$

$$P(k) = \frac{P(k-1)}{\lambda}[I - K(k)\varphi(k)^T], \tag{6.7}$$

where the parameter vector is

$$\hat{\theta}^T = [\hat{a}_1 \quad \hat{a}_2 \quad \hat{b}_0 \quad \hat{b}_1]$$

and the regressor vector is

$$\hat{\varphi}(k-1)^T = [-F_p(k-1) \quad -F_p(k-2) \quad f_c(k-1) \quad f_c(k-2)].$$

$K(k)$ is called the estimation gain with a user-selected forgetting factor of $0.5 < \lambda < 1$. Typically λ is selected between 0.8 and 0.95 to discount the influence of previous measurements on the estimation. The covariance matrix $P(k)$ is a square matrix with dimensions of $N_p \times N_p$, where N_p is the number of parameters to be estimated (i.e., $N_p = 4$ in this application). The algorithm is started by assigning guessed initial values for the parameters (i.e., $\theta(0) = [0.1\ 0.1\ 0.1\ 0.1]$) and a large initial value for the covariance matrix ($P(0) = 10^5$). Care must be taken in using the RLS algorithm. When the adaptive control algorithm runs for a long period and the process does not change significantly, the covariance matrix becomes either too small or too large, which creates numerical instability. The problem can be avoided by monitoring the trace of the covariance matrix (i.e., $\text{tr}(P(k)) = \sum_{i=1}^{4} P_{ii}$). Whenever the trace is too small or too large, the covariance matrix can be reset to an initial value.

6.3.3 Pole-Placement Control Algorithm

Once the process parameters are estimated on-line, a regular pole-placement control algorithm [86] can be designed to constrain the cutting forces (see Fig. 6.5). The closed-loop transfer function of the control system is given by

$$\frac{F_p(k)}{F_r(k)} = \frac{T(z)B(z)}{A(z)R(z) + B(z)S(z)}. \tag{6.8}$$

The polynomials $S(z^{-1})$, $T(z^{-1})$, and $R(z^{-1})$ represent feedback, feedforward, and error regulators respectively, which have to be determined adaptively at each spindle period. The control algorithm regulates the feeding velocity $f_c(k)$ as

$$R\, f_c(k) = T F_r(k) - S\, F_p(k). \tag{6.9}$$

The aim of the adaptive controller based on pole placement design is to make the closed-loop transfer function between the reference force F_r and the actual

Figure 6.5: Block diagram of an adaptive pole placement control system in machining.

force F_p obey the following desired model dynamics:

$$\frac{F_p(k)}{F_r(k)} = \frac{B_m(z)}{A_m(z)}. \tag{6.10}$$

Here $A_m(z)$ is designed to satisfy the transient response characteristics of the controller. A second-order dynamics is selected to represent the desired response:

$$A_m(z) = z^2 - 2e^{-\zeta\omega_n T} \cos(\omega_n \sqrt{1-\zeta^2}T)z + e^{-2\zeta\omega_n T}. \tag{6.11}$$

This corresponds to a second-order continuous system with a desired damping ratio of ζ and natural frequency of ω_n (i.e., $s^2 + 2\zeta\omega_n s + \omega_n^2$) with a discrete control interval of T, which is equal to the spindle period. A damping ratio of $\zeta = 0.8$ and a rise time (T_r) equal to the larger of three spindle periods or feed drive servo's rise time can be selected for practical applications. The natural frequency corresponds to $\omega_n = 2.5/(T_r)$ rad/s. With these criteria the discrete closed-loop model transfer function with a unit gain is

$$\frac{F_p(k)}{F_r(k)} = \frac{B_m}{A_m} = \frac{1 + m_1 + m_2}{z^2 + m_1 z + m_2}, \tag{6.12}$$

where $m_1 = -2e^{-\zeta\omega_n T} \cos(\omega_n \sqrt{1-\zeta^2}T)$ and $m_2 = +e^{-2\zeta\omega_n T}$. For the system to behave like the desired model, the following equivalence must hold:

$$\frac{BT}{AR + BS} = \frac{B_m}{A_m}. \tag{6.13}$$

The design problem is to find the polynomials R, S, and T using the machining plant's and desired model's transfer functions. The numerator of the machining process transfer function can be expressed as follows:

$$B(z) = B^-(z)B^+(z) = b_0(z + b_1/b_0). \tag{6.14}$$

In reality, the zero of the machine tool plant is always stable and within the unit circle (i.e., $|b_1/b_0| < 1.0$). However, RLS estimation does not necessarily predict the parameters b_0 and b_1 accurately. To keep the control system robust, separate control systems are designed depending on whether the process has a stable or an unstable zero.

Case i: Process Zero Is Stable ($|b_1/b_0| < 1.0$). Let

$$B^-(z) = b_0, \quad B^+(z) = z + b_1/b_0.$$

The control parameters can be designed in such a way that the stable zero can be canceled by one of the system's closed-loop poles. To cancel B^+ with one of the closed-loop transfer function poles, as A and B are coprime, R must be designed to have a factor of B^+:

$$R = B^+ \cdot R' = (z + b_1/b_0) \cdot R'. \tag{6.15}$$

The closed-loop transfer function (Eq. 6.13) becomes

$$\frac{B^+ B^- T}{B^+(AR' + B^- S)} = \frac{B^- T}{AR' + B^- S} = \frac{A_0 \cdot B_m(z)}{A_0 \cdot A_m(z)}, \tag{6.16}$$

where the observer A_0 is included to ensure good tracking and causality in the design. (Note: Causality means that the control input does not depend on the future inputs and outputs of the system.) The control polynomials R', S, and T can be solved from the following identities:

$$\begin{aligned} B^- T &\equiv A_0 B_m, \\ AR' + B^- S &\equiv A_0 A_m, \end{aligned} \tag{6.17}$$

where the second identity is called the *Diophantine Equation*. The degrees of machining process transfer functions (i.e., the order of polynomials B and A) are

$$\deg(B) = \deg(B^+) + \deg(B^-) = 1 + 0, \quad \deg(A) = 2 \tag{6.18}$$

and $\deg(A_m) = 2$. The *degree* of the polynomials are determined from the causality conditions and the order of process transfer functions and are summarized as follows:

$$\begin{aligned} \deg(A_0) &\geq 2 \cdot \deg(A) - \deg(A_m) - \deg(B^+) - 1 = 2 \times 2 - 2 - 1 - 1 = 0, \\ \deg(R) &= \deg(A_0) + \deg(A_m) - \deg(A) = 1 + 2 - 2 + 1 = 2, \\ \deg(S) &= \deg(A) - 1 = 2 - 1 = 1, \\ \deg(T) &= \deg(A_0) + \deg(B_m) - \deg(B^-) = 1 + 0 - 0 = 1. \end{aligned}$$

The resulting control polynomials are

$$\begin{aligned} A_0(z) &= z, \\ R'(z) &= z + r_1, \\ S(z) &= s_0 z + s_1, \\ T(z) &= t_0 z. \end{aligned} \tag{6.19}$$

The Diophantine equation ($AR' + B^- S = A_0 A_m$) becomes

$$(z^2 + a_1 z + a_2)(z + r_1) + b_0(s_0 z + s_1) \equiv z(z^2 + m_1 z + m_2)$$

and

$$B^- T \equiv A_0 B_m.$$

From the two equivalence, the control polynomial parameters are derived as follows:

$$r_1 = m_1 - a_1, \quad \rightarrow \qquad\qquad R(z) = z^2 + (r_1 + b_1/b_0)z + (b_1/b_0)r_1,$$

$$s_0 = \frac{m_2 - a_1 r_1 - a_2}{b_0}, \quad s_1 = -\frac{a_2 r_1}{b_0}, \quad \rightarrow \quad S(z) = s_0 z + s_1,$$

$$t_0 = \frac{1 + m_1 + m_2}{b_0}, \quad \rightarrow \qquad\qquad T(z) = t_0 z,$$

(6.20)

where $R(z) = B^+ R' = (z + b_1/b_0)(z + r_1)$.

The feed rate is calculated at each spindle period from the control law,

$$f_c(k) = \frac{T(z)}{R(z)} F_r(k) - \frac{S(z)}{R(z)} F_p(k).$$

After substituting parameters and scaling polynomials by z^{-2}, we obtain

$$f_c(k) = t_0 F_r(k-1) - \left(r_1 + \frac{b_1}{b_0}\right) f_c(k-1) - r_1 \frac{b_1}{b_0} f_c(k-2)$$

$$- s_0 F_p(k-1) - s_1 F_p(k-2).$$

(6.21)

Here, the process parameters are replaced by the on-line estimated values $(\hat{b}_0, \hat{b}_1, \hat{a}_1, \hat{a}_1)$ so that the control system can adapt itself to the time-varying machining process. The real time application of the adaptive control algorithm can be summarized as follows:

- Collect the maximum cutting force at spindle revolution k.
- Estimate the process parameters (b_0, b_1, a_1, a_2) using the RLS algorithm.
- Solve for the control polynomials R, S, and T (Eq. 6.20).
- Calculate the feed rate (Eq. 6.21) and send it to the CNC.

Case ii: Process Zero Is not Stable ($|b_1/b_0| > 1.0$). Let

$$B^+(z) = 1, \quad B^-(z) = b_0 z + b_1.$$

The procedure is identical to that of case i except that the process zero is not canceled and it must be included in the model transfer function:

$$t_0 = \frac{1 + m_1 + m_2}{b_0 + b_1},$$

and the Diophantine equation becomes

$$(z^2 + a_1 z + a_2)(z + r_1) + (b_0 z + b_1)(s_0 z + s_1) \equiv z(z^2 + m_1 z + m_2).$$

The remaining solution procedure is identical to the previous case.

Machining Test Results. The adaptive pole-placement control algorithm is integrated to the Intelligent Machining Module as a Plug In Module, and tested

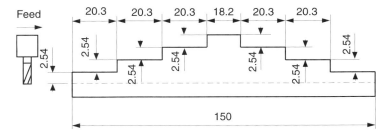

Figure 6.6: Workpiece geometry to test step response of adaptive milling force control algorithms. Workpiece material: Aluminum 6067. Cutter: 4-fluted carbide end mill. Spindle speed = 715 rev/min.

on a milling machine controlled by an open architecture CNC. The workpiece geometry and cutting conditions are given in Figure 6.6. The workpiece was prepared to have step changes in the axial depth of cut in order to test the transient response and robustness of the control algorithm. Whenever an unstable zero is estimated by the RLS algorithm, the pole-placement control law is switched to the second case as outlined above. The reference peak force was set to $F_r = 1,200$ N. The experimental results are given in Figure 6.7. The peak resultant force is always kept at the desired 1,200 N level, and the algorithm gives some overshoots when the axial depth of cut increases stepwise. However, the control algorithm is stable and rapidly converges. The corresponding feeds identified by the control law are also given, along with the estimated process parameters (b_0, b_1, a_1, a_2) by the RLS PIM. The control interval corresponds to the spindle rotation period.

6.3.4 Adaptive Generalized Predictive Control of Milling Process

In general, most adaptive control algorithms work fairly well when they are tuned for a specific machine tool at a constant speed, such as the adaptive pole-placement control presented above. The pole-placement algorithm depends on the spindle speed, which is the sampling period (T). A robust and modular adaptive control algorithm, which is spindle speed independent, free of covariance drifting in parameter estimation, and robust to variable time delays between the feed command and actual feed delivered by the CNC is required so that it can be integrated to an open-architecture CNC system with a minimal amount of modifications. Based on the extensive experience of machining process control strategies in the author's laboratory [89, 91], an adaptive Generalized Predictive Control (GPC) method was found to satisfy the requirements most [88]. GPC is more robust to future transient changes in the cutting forces, i.e., the geometry which can be obtained from CAD system [89], and the varying time delay between the feed command generated by the adaptive control module and its actual execution by the CNC system. The fundamental control law was presented by Clarke et al. [92], and its specific application to machining control is briefly summarized as follows.

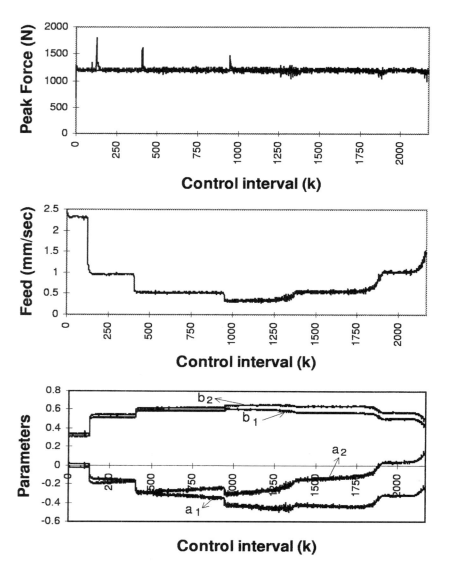

Figure 6.7: Milling test results of adaptive pole-placement control system. See Figure 6.6 for the part geometry and cutting conditions.

The combined transfer function of the feed drive dynamics and chatter-free cutting process can be expressed as [88]

$$G_c(z^{-1}) = \frac{F_p(k)}{f_c(k)} = \frac{z^{-1}B(z^{-1})}{A(z^{-1})}$$
$$= \frac{z^{-1}(b_0 + b_1 z^{-1} + b_2 z^{-2})}{1 + a_1 z^{-1} + a_2 z^{-2}}, \tag{6.22}$$

where $f_c(k)$ is the command feed sent to the CNC system by the adaptive control module at spindle period k. With respect to the previous transfer function

(Eq. 6.4), the order of the machining process is increased to account for the non-linear relationship between the feed rate (i.e., chip thickness) and the cutting force. The parameters of polynomials B and A may vary with time depending on the changes in the radial and axial depths of cut and the slight changes in the feed drive dynamics. They are estimated recursively at each spindle or adaptive control interval k from measured peak force F_p and commanded feed f_c vectors. A modified recursive least squares (RLS) algorithm, which avoids parameter drifting caused by the absence of steady excitations (i.e., changes in the work-piece geometry), was used. The covariance matrix drifting toward instability is avoided by tracking the trace of the covariance matrix, and updating it only when there is a change in the process. The recursive parameter identification algorithm with covariance tracking is given in Appendix B. The GPC method is devised for an ARIMAX model. Invoking this model naturally leads to the adoption of an integrator to the controller thus eliminating steady state offsets. The machining process model is therefore rearranged in the ARIMAX form:

$$F_p(k) = \frac{B(z^{-1})}{A(z^{-1})} f_c(k-1) + \frac{\zeta(k)}{\Delta A(z^{-1})}, \tag{6.23}$$

where $\Delta = (1 - z^{-1})$ and $\zeta(k)$ is assumed to be an uncorrelated random noise sequence. Then a j step ahead prediction of $F_p(k)$ based on Eq. (6.23) can be obtained by expressing the noise term by its partial fraction expansion,

$$\frac{z^j \zeta(k)}{\Delta A(z^{-1})} = z^j E_j \zeta(k) + \frac{F_j(k)}{\Delta A(z^{-1})} \zeta(k),$$

which leads to the Diophantine equation,

$$1 = E_j(z^{-1})A(z^{-1})\Delta + z^{-j}F_j(z^{-1}), \tag{6.24}$$
$$\deg(E_j(z^{-1})) = j - 1 , \quad \deg(F_j(z^{-1})) = \deg(A(z^{-1})) = 2.$$

Here E_j and F_j are polynomials uniquely defined, given $A(z^{-1})$ and the prediction interval j. Once they are calculated for one value of j, simple iterations can be used to calculate them for all other j, see the following section. All the noise components are in the future; therefore provided that the output data up to time k and $f_c(k + j - 1)$ are available, the j step ahead prediction of peak cutting force F_p at spindle interval k can be given by

$$\hat{F}_p(k + j) = G_j(z^{-1})\Delta f_c(k + j - 1) + F_j F_p(k), \tag{6.25}$$

where $G_j(z^{-1}) = E_j B(z^{-1})$. The delay in the process is assumed to be one here, which may change slightly depending on the spindle speed.

A general case is considered by selecting the minimum and maximum prediction output horizons as $N_1 = 1$ and $N_2 = 4$, respectively. The GPC control law takes predicted output values within the next four spindle revolutions when manipulating the control parameter (i.e., the feed). Equation (6.25) contains present (i.e., k), past (i.e., $k - i$), and future (i.e., $k + i$) values of feed and forces. GPC considers that future control inputs (i.e., feed rates $f_c(k+j)$) do not change beyond the control horizon NU, which is selected to be $NU = 1$ here. Thus, $\Delta f_c(k+1) = \Delta f_c(k+2) = \Delta f_c(k+3) = 0$. Equation (6.25) can be partitioned as

$$\{\hat{F}_p\} = \{G_I\}\Delta f_c(k) + \{f\}, \tag{6.26}$$

where vector dimensions are $[4 \times 1]$, and their derived contents are given in the following section. Note that vector $\{f\}$ contains present and past measured peak forces $F_p(k-i), i = 0, 1, 2$ and past feed commands $f_c(k-i), i = 1, 2$ as well as recursively computed polynomials G_j and F_j. According to GPC strategy, the feed rate is calculated by minimizing the expected value of a quadratic cost function containing future predicted errors within the output horizon (i.e., the next four spindle revolutions) between the actual and reference peak forces over the control horizon (i.e., the next spindle revolution):

$$J(f_c, k) = E \sum_{j=N_1}^{N_2} [\hat{F}_p(k+j) - F_r(k+j)]^2 + \lambda[\Delta f_c(k)]^2, \tag{6.27}$$

where λ, the weighting factor on the control input increment $f_c(k) - f_c(k-1)$, is selected as $\lambda = 0.2$ to soften the impact of sudden changes in the geometry on the adaptive control law. The minimization of the cost function ($\partial J/\partial f_c = 0$) leads to the input feed command at spindle interval k:

$$f_c(k) = f_c(k-1) + \frac{1}{\{G_I\}^T\{G_I\} + \lambda}\{G_I\}^T(\{F_r\} - \{f\}). \tag{6.28}$$

The parameters of polynomials E_j, F_j, and G_j are recursively calculated followed by the control law given above (Eq. 6.28) at each spindle period. The computed vector feed $f_c(k)$ is sent to the CNC master controller, which constrains the actual machining force at a desired reference level. The feed is bounded between the user-defined minimum and maximum limits for safety.

Recursive Computation of Polynomials E_j, F_j, and G_j

The recursive algorithm of GPC is briefly summarized here. For $j = 1$ the Diophantine equation becomes

$$1 = \Delta A(z^{-1})E_1(z^{-1}) + z^{-1}F_1(z^{-1}), \tag{6.29}$$

which yields $e_{10} = 1$, $f_{10} = -(a_1 - 1)$, $f_{11} = -(a_2 - a_1)$, and $f_{12} = a_2$.

For $j > 1$ consider the difference of Diaphontine equations at j and $j + 1$:

$$\Delta A[E_{j+1} - E_j] + z^{-j}[z^{-1}F_{j+1} - F_j] = 0,$$

which leads to

$$e_{10} = e_{20} = e_{30} = e_{40} = e_0 = 1; \quad e_{21} = e_{31} = e_{41} = e_1; \quad e_{32} = e_{42} = e_2;$$
$$e_{43} = e_3;$$

and

$$e_{j-1} = f_{j-1,0}; \quad f_{j,i} = f_{j-1,i+1} - e_{j-1}(a_{i+1} - a_i), \quad i = 0, 1, 2.$$

TABLE 6.1. Polynomials E_j, F_j, and G_j

j	E_j	F_j	G_j
1	e_0	$f_{10} + f_{11}z^{-1} + f_{12}z^{-2}$	$g_{10} + g_{11}z^{-1} + g_{12}z^{-2}$
2	$e_0 + e_1z^{-1}$	$f_{20} + f_{21}z^{-1} + f_{22}z^{-2}$	$g_{20} + g_{21}z^{-1} + g_{22}z^{-2} + g_{23}z^{-3}$
3	$e_0 + e_1z^{-1} + e_2z^{-2}$	$f_{30} + f_{31}z^{-1} + f_{32}z^{-2}$	$g_{30} + g_{31}z^{-1} + g_{32}z^{-2} + g_{33}z^{-3}$ $+ g_{34}z^{-4}$
4	$e_0 + e_1z^{-1} + e_2z^{-2} + e_3z^{-3}$	$f_{40} + f_{41}z^{-1} + f_{42}z^{-2}$	$g_{40} + g_{41}z^{-1} + g_{42}z^{-2} + g_{43}z^{-3}$ $+ g_{44}z^{-4} + g_{45}z^{-5}$

The polynomial G_j is then calculated:

$$G_j = E_j B = E_j(b_0 + b_1z^{-1} + b_2z^{-2}). \tag{6.30}$$

The equations are solved recursively at each prediction step j. The results are given in Tables 6.1 and 6.2.

The contents of the vectors are given as

$$\{\hat{F}_p\} = \begin{Bmatrix} \hat{F}_p(k+1) \\ \hat{F}_p(k+2) \\ \hat{F}_p(k+3) \\ \hat{F}_p(k+4) \end{Bmatrix}, \quad \{F_r\} = \begin{Bmatrix} F_r(k+1) \\ F_r(k+2) \\ F_r(k+3) \\ F_r(k+4) \end{Bmatrix}, \quad \{G_I\} = \begin{Bmatrix} g_0 \\ g_1 \\ g_2 \\ g_3 \end{Bmatrix}, \tag{6.31}$$

$$\{f\} = \begin{Bmatrix} g_{11} & g_{12} & f_{10} & f_{11} & f_{12} \\ g_{22} & g_{23} & f_{20} & f_{21} & f_{22} \\ g_{33} & g_{34} & f_{30} & f_{31} & f_{32} \\ g_{44} & g_{45} & f_{40} & f_{41} & f_{42} \end{Bmatrix} \begin{Bmatrix} \Delta f_c(k-1) \\ \Delta f_c(k-2) \\ F_p(k) \\ F_p(k-1) \\ F_p(k-2) \end{Bmatrix}. \tag{6.32}$$

Machining Test Results. The adaptive GPC algorithm is also integrated to the Intelligent Machining Module as a Plug In Module and tested on a milling machine controlled by an open-architecture CNC. The workpiece geometry and cutting conditions are given in Figure 6.6, which is the same as the pole-placement

TABLE 6.2. Recursively Calculated Parameters of Polynomials E_j, F_j, and G_j

j	Parameters
1	$e_0 = 1$; $f_{10} = -(a_1 - 1)$, $f_{11} = -(a_2 - a_1)$, $f_{12} = a_2$; $g_0 = g_{10} = b_0$, $g_{11} = b_1$, $g_{12} = b_2$
2	$e_1 = f_{10}$; $f_{20} = f_{11} - e_1(a_1 - 1)$, $f_{21} = f_{12} - e_1(a_2 - a_1)$, $f_{22} = e_1a_2$; $g_{20} = b_0$, $g_1 = g_{21} = b_1 + e_1b_0$, $g_{22} = b_2 + e_1b_1$, $g_{23} = e_1b_2$
3	$e_2 = f_{20}$; $f_{30} = f_{21} - e_2(a_1 - 1)$, $f_{31} = f_{22} - e_2(a_2 - a_1)$, $f_{32} = e_2a_2$; $g_{30} = g_0$, $g_{31} = g_1$, $g_2 = g_{32} = b_2 + e_1b_1 + e_2b_0$, $g_{33} = e_1b_2 + e_2b_1$, $g_{34} = e_2b_2$
4	$e_3 = f_{30}$; $f_{40} = f_{31} - e_3(a_1 - 1)$, $f_{41} = f_{32} - e_3(a_2 - a_1)$, $f_{42} = e_3a_2$; $g_{40} = g_0$, $g_{41} = g_1$, $g_{42} = g_2$, $g_3 = g_{43} = e_1b_2 + e_2b_1 + e_3b_0$, $g_{44} = e_2b_2 + e_3b_1$, $g_{45} = e_3b_2$

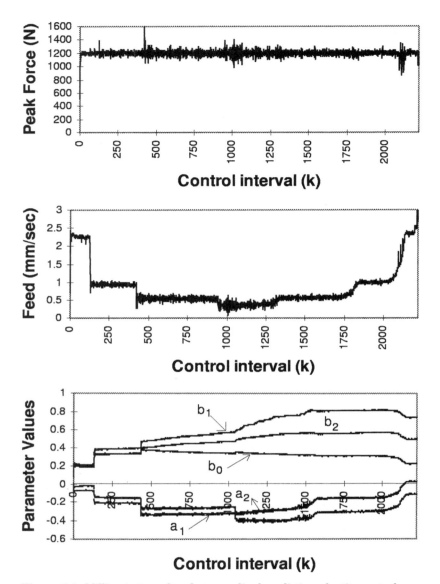

Figure 6.8: Milling test results of a generalized predictive adaptive control system. See Figure 6.6 for the part geometry and cutting conditions.

control tests shown in Figure 6.7. The reference peak force was set to $F_r = 1,200$ N. The experimental results are given in Figure 6.8. The peak resultant force is always kept at the desired 1,200 N level, and the algorithm gives some overshoots when the axial depth of cut increases stepwise. However, the control algorithm is stable and rapidly converges. The corresponding feeds identified by the control law are also given, along with the estimated process parameters (b_0, b_1, b_2, a_1, a_2) by the RLS PIM. The control interval corresponds to the spindle rotation period.

6.3.5 In-Process Detection of Tool Breakage

The tool breakage detection algorithm uses the average resultant cutting forces per tooth period (m), which can be written as:

$$F_a(m) = \frac{\sum_{i=1}^{I}\sqrt{F_x(i)^2 + F_y(i)^2}}{I},\tag{6.33}$$

where I is the number of force samples collected at tooth period m. When run-out and tooth breakage are not present, and the cutter is not in a transient geometric zone (i.e., the part–cutter intersection geometry is not changing), all teeth on the milling cutter produce equal average cutting forces, and the first differences of the average cutting forces,

$$\Delta F_a(m) = F_a(m) - F_a(m-1) = (1 - z^{-1})F_a(m),\tag{6.34}$$

will be zero. Otherwise, the average cutting forces will reflect the changes in the chip load and be nonzero. If the cutter runs into a transient geometry (i.e., varying entry and exit angles due to holes, slots, and voids), the differenced forces will reflect the trend [93]. Since the feed per tooth period is considerably smaller than the diameter of the cutter and the transient geometry changes, a first-order adaptive time series filter can remove the slow varying dc trend caused by the changes in the workpiece geometry:

$$\epsilon_1(m) = (1 - \hat{\phi}_1 z^{-1})\Delta F_a(m),\tag{6.35}$$

where $\hat{\phi}_1$ is estimated from measurements $\Delta F_a(m)$ using the standard recursive least squares technique at each tooth period. However, such a filter may still produce high amplitude residuals [94] at each tooth period if the run-outs on the cutter teeth are not the same. The run-out of each tooth can be removed by comparing the tooth's performance by itself one revolution before:

$$\Delta^N F_a(m) = F_a(m) - F_a(m-N).\tag{6.36}$$

The resulting differences are again passed from a first-order adaptive time series filter to remove the possible dc trend left by the changes in the geometry:

$$\epsilon_2(m) = (1 - \hat{\phi}_2 z^{-1})\Delta^N F_a(m).\tag{6.37}$$

The two adaptive time series filters are run recursively in parallel at every tooth period. When the cutting forces increase from a level of air machining at the beginning of cut, the maximum residuals of both filters are measured during the first five revolutions of spindle. It is assumed that the cutter is not broken during this period. The residuals contain the influence of run-out on the cutter used and the noise, but not the geometric transients, which are filtered by the algorithm. The breakage thresholds are selected by scaling the maximum residuals by user-defined factors α_1 and α_2,

$$LIMIT_1 = \alpha_1 \cdot \max\{\epsilon_1\}, \quad LIMIT_2 = \alpha_2 \cdot \max\{\epsilon_2\},\tag{6.38}$$

and they are used during the rest of the machining with the cutter. A tool breakage event is assumed whenever both residuals exceed their thresholds,

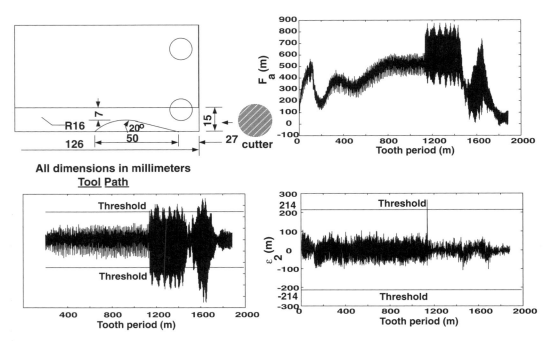

Figure 6.9: Typical tool breakage event. One flute has a chipped edge at tooth period 1138. Cutting conditions: 25.4 mm diameter HSS cutter with 4 flutes, spindle speed = 566 rev/min, feed rate = 0.1 mm/tooth, depth of cut = 7 .62 mm; radial width of cut is varying.

$\epsilon_1(m) > LIMIT_1$ and $\epsilon_2(m) > LIMIT_2$ followed by an additional transient assurance check in the following revolution $\epsilon_1 > LIMIT_1$.

Figure 6.9 illustrates a sample testing of the algorithm for peripheral milling of a workpiece with transient voids and holes. The cutting test was repeated twice under identical cutting conditions, but one with a good cutter and the other with a chipped end mill. Later, the collected average force data are assembled at various transient geometry locations to test the robustness of the algorithm. The measured force and processed residuals are also shown in Figure 6.9. The intuitively selected threshold factors were $\alpha_1 = \alpha_2 = 2.0$, which correspond to $LIMIT_1 = 245$ N and $LIMIT_2 = 214$ N. The tool failure is detected since both residuals exceed their thresholds at the tool breakage event ($m = 1,138$), and ϵ_1 continuously exceeds its threshold thereafter without forgetting the breakage event.

The same algorithm can be applied to any sensor measurements provided that the signals can be correlated to cutting forces at the operating frequency (i.e., bandwidth) of the machine tool. For example, the feed drive motor current has been used at low cutting speeds for tool condition monitoring [73].

6.3.6 Chatter Detection and Suppression

The chatter is detected by continuously monitoring the amplitude of the sound spectrum measured by a microphone. It was shown previously that variation of spindle speed slightly improves the stability in milling, but limited spindle torque/power and frequency bandwidth prevents practical application of the method [95]. Smith and Tlusty [96] suppress the chatter by matching the tooth passing frequency with the on-line detected chatter frequency. This results in the operation of milling at the highest stability lobe, which is the most favorable speed zone for chatter-free machining. Alternatively, Weck [97] proposed a method of automatically reducing the axial depth of cut for chatter suppression. The reduction in the axial or radial depth of cut decreases the gain of the dynamic milling operation, which has a closed-loop behavior. Although the later strategy lowers the productivity, it always works and is especially applicable in the low spindle speed range. Since chatter occurs close to one of the structural frequencies, which are generally higher than 500 Hz, high spindle

Figure 6.10: Spectrum of sound during stable and unstable machining. Cutting conditions: 19.05 mm diameter HSS cutter with 4 flutes, spindle speed = 885 rev/min, feed rate = 0.01 mm/tooth, helix angle = 0, tool overhang = 74.0 mm, half immersion upmilling, workpiece = Al 7075-T6 alloy.

speed is required to avoid chatter by matching the tooth and chatter frequencies. Here the power spectrum of the sound is calculated, and the maximum amplitude that occurs at the chatter frequency is searched every 250 ms. A sample spectrum of milling operation with and without chatter vibrations are shown in Figure 6.10. The magnitude of spectrum at the chatter frequency of 2,300 Hz is about six times larger than the one in chatter-free milling. The cutting conditions and the microphone location were identical, except that the depth of cut was increased from 2.54 mm to 6.35 mm in the unstable milling with chatter vibrations. It was observed that once the microphone is set for a particular machine, the sound spectrum does not change significantly with the changes in the feed rate and axial and radial depths of cut when there are no chatter vibrations. The chatter threshold is then selected as 5 mV, which is well above the maximum amplitude observed in stable, chatter-free machining tests. The CNC system assumes that chatter is present whenever the measured sound spectrum amplitude exceeds the threshold (5 mV) during milling on the particular machine tool setup [88].

6.4 EXAMPLE: INTELLIGENT POCKETING WITH THE IMM SYSTEM

The generalized predictive control methods of cutting forces, tool breakage detection, and chatter avoidance algorithms explained in the previous sections were integrated to our open-architecture CNC/IMM system. All three algorithms run in parallel and are able to manipulate the CNC system. The complete system was tested on a pocket milling application (see Fig. 6.11). The full axial depth of cut was programmed in the NC code sent to the CNC system. Default minimum and maximum feed rates are automatically assumed whenever the adaptive force control job is envoked by the user. The adaptive control manipulates the feed within the feed rate range during machining. The cutter fully penetrates to the bottom of the pocket at path position 1 and starts finishing the wall as it moves along the x axis toward 2. The wall's dimensional tolerance is 0.1 mm, and the cutter stiffness at the tip was 2,700 N/mm. The reference force to be controlled is selected to be the maximum force normal to the wall (F_y) and is set to 270 N to constrain the static deflection of the end mill within the tolerance. The system detects the chatter, stops the feed, decreases the axial depth of cut by 1 mm, and continues to cut with the feed regulated by the adaptive controller. The system automatically reduced the depth of cut from 6.35 mm to 3.35 mm where the chatter diminished (see Fig. 6.11). The adaptive control system adjusted the feed to 74.3 mm/min. When the tool moves from 2 to 3, the normal force to the wall is F_x, which is kept at 270 N. The same algorithm is repeated. If the chatter had occurred, the depth of cut would have been further reduced from 3.35 mm. The procedure is repeated at paths 3–4 and 4–5. The minimum depth of cut was assumed to be removed during the first layer, the NC program was updated, and the remaining wall was finished in one more pass, automatically. The cutter was positioned to path point 6, and the remaining island was removed with a reference resultant force set to 600 N to avoid shank failure. The tool breakage monitoring was always active during machining, and the algorithm never indicated a breakage event (see Fig. 6.12). Figure 6.13 shows the sound spectrum for machining the pocket. After the machining of the pocket, the walls had a 0.095 mm maximum form error, which was within the tolerance specified.

Other process control algorithms, such as thermal deformation compensation, collision detection, and in-process probing, can be added to the system by simply registering the new PIMs to the IMM library.

6.5 PROBLEMS

1. Design an adaptive pole-placement control system to keep cutting forces at a desired reference level in a turning process. The spindle speed is kept constant at 1,200 rev/min, and the resultant cutting force constant is given as $K_c = 1,500$ N/mm^2. The axial depth of cut is increased by 1.0 mm every 50 revolutions of the spindle, starting with 1.0 mm initial depth. When the spindle completes 250 revolutions, the axial depth of cut is decreased at the same rate. The machine tool's feed control servo dynamics can be approximated as a first-order continuous system with a time constant of $\tau_m = 0.1$ s. The vibration-free cutting process has an approximate time

Figure 6.11: Machining a pocket. Cutting conditions: 15.875 mm diameter HSS cutter with 2 flutes, spindle speed = 1,500 rev/min, varying feed rate, helix angle = 30 degrees, tool overhang = 76.2 mm, workpiece = Al 7075-T6 alloy.

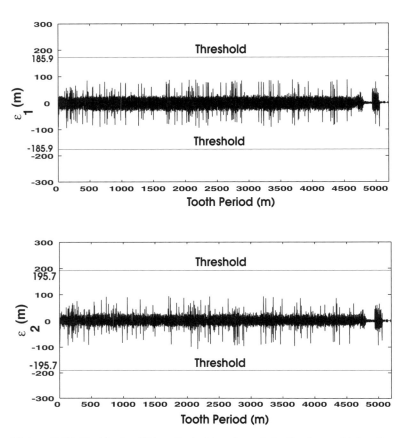

Figure 6.12: Residuals of filters for tool breakage detection in machining the pocket.

Figure 6.13: Maximum amplitude of sound spectrum during machining the pocket. (Note: Only a part of the first pass of machining is shown in the figure.)

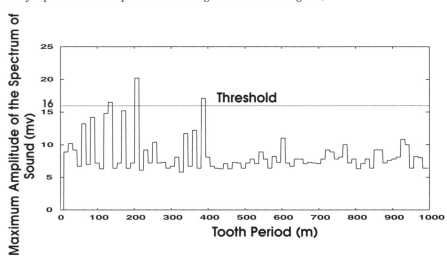

constant of $\tau_c = 0.65\,T$, where T is the spindle period. The adaptive control system must behave like a second-order system with a damping ratio of 0.95 and rise time of 4 spindle revolutions.

2. Design an adaptive Proportional Integral Darivative (PID) controller for the same process given in Problem 1.

3. Design an adaptive GPC for the same process given in Problem 1.

4. How can you modify the adaptive force controller so that it keeps the cutter deflections within 0.2 mm? Discuss the practical aspects of the application in minimizing the surface errors caused by the static deflections of the cutter.

5. Write a tool breakage detection program in the C language. Test the algorithm by simulating a tool breakage event within the milling force prediction algorithm.

6. How can you detect tool breakage by monitoring the power spectrum of cutting forces?

7. A piezo actuator–based tool holder is designed for precision positioning of cutting tools during turning operations. The picture of the piezo actuator–based tool holder mounted on the turret of a CNC turning center and its mechanical and control diagram are shown in Figure 6.14. One end of the piezo element rests on a solid, rigid wall, while the other end is bolted to a flexure mechanism carrying a tool holder. A signal voltage (u(V)) is supplied by the computer to the piezo electric high voltage amplifier, which delivers a very high negative voltage (v(V)). When a high voltage is applied to the piezo element, it pushes or pulls the flexure carrying the tool holder. The displacement of the tool holder is measured by a laser. Using a digital control system, it is possible to control the position of the tool tip within 0.1 m accuracy. The actuator is used within a frequency bandwidth of 700 Hz and thus any dynamics above 1,000 Hz can be neglected. The frequency response behavior of each component in the system is shown in Figure 6.15. The tool holder can be modeled as an equivalent mass (m), spring (k), and damping (c) elements as shown in the figure. Design a control system by strictly following the order of instructions outlined below:

    (a) Draw the block diagram of a computer control system for the precision positioning of the tool tip.

    (b) Identify the transfer function of each physical component in the block diagram. Use symbols, but summarize the numerical values of each symbol at the end.

    (c) Assuming that the control/sampling interval is T seconds, express open-loop continuous and discrete (zero-order hold equivalent) transfer functions of the entire physical system algebraically. (Don't use numbers.)

Photograph of the piezo actuator assembly.

Piezo actuator mounted on CNC turning center turret.

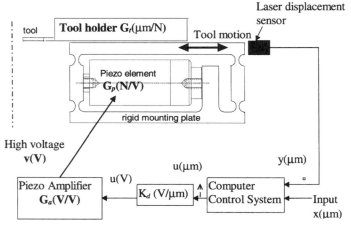

Figure 6.14 A piezo actutor–based precision tool holder system mounted on a CNC turning center turret. See Problem 7 for details.

(d) Draw the block diagram of a pole-placement control system for ultra-positioning of the tool tip.

(e) Design a pole-placement control law in such a way that the settling time and overshoot are t_s, M_p, respectively. (Find the corresponding desired natural frequency (w_m) and damping ratio (η) first, and use these two symbols in your algebraic derivations.) Solve the design problem analytically using only symbolic algebra.

(f) The following numerical values are provided: $T = 0.0004$ s, $t_s = 0.005$ s, $M_p = 1\%$. Tool holder: $m = 2.1$ kg, $c = 0.0006324$ N/(m/s), $k = 33.55$ N/m. D/A converter (K_d): 16 bit, -3V to $+3$V range. Summarize the numerical values of all unknowns in the following order:

  • G_a, G_p, G_t: Transfer functions of piezo amplifier, piezo element, and tool holder.

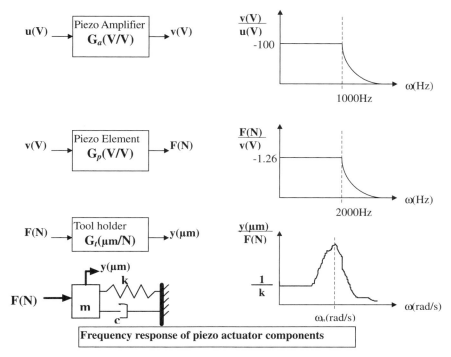

Figure 6.15 Frequency response functions of the piezo actuator assembly.

- $G_o(s)$, $G_o(z)$: Open-loop transfer functions of the entire physical piezo actuator assembly.
- B_m/A_m: Desired transfer function in z domain.
- R, S, T: Control law parameters.
- $u(k) = f(x, u, y)$: Expression for control input.

LAPLACE AND z TRANSFORMS

A.1 INTRODUCTION

The Laplace and z transform methods are useful for simplifying the analysis and design of linear time-invariant systems in continuous time and in discrete time, respectively. For example, considering a set of differential equations representing a system in continuous time, Laplace transform theory simplifies the problem by replacing the given set of differential equations by a set of algebraic equations.

In this appendix we will consider only linear time-invariant dynamic systems. For a more detailed study of the Laplace, z, and related transforms such as the Fourier and starred Laplace transforms and their interrelationships refer to established control texts (e.g., Ogata [74, 75] and Kuo [98]).

The use of Laplace (s) and z transforms are sometimes confused by engineers who have limited experience in control theory. The following simple example is first introduced to illustrate the physical meaning and the use of both transformations in engineering.

Consider a simple first-order system (i.e., a servo motor) whose input is x (i.e., voltage given to the amplifier) and output is y (i.e., motor velocity). The system transfer function in the Laplace (i.e., continuous) domain is:

$$G(s) = \frac{Y(s)}{X(s)} = \frac{K}{\tau s + 1} = K \frac{1/\tau}{s + 1/\tau}, \tag{A.1}$$

where K is the gain (at $s = 0$ or $t = +\infty$) and τ is the time constant of the motor. If a step input with an amplitude of A (i.e., physically A volts) is given to the system, we want to know what happens to the output y (i.e., motor velocity [rad/s]). Considering that the Laplace transform of step input is $L[x(t)] = X(s) = A/s$, we have

$$Y(s) = X(s) \cdot K \frac{1/\tau}{s + 1/\tau} = KA \frac{1/\tau}{s(s + 1/\tau)}. \tag{A.2}$$

The inverse Laplace transform of Eq. (A.2) provides the response (i.e., velocity) of the system (motor) for a step input (i.e., step change in the supply voltage) with an amplitude of A [in volts] in the continuous time domain:

$$y(t) = L^{-1} Y(s) = KA(1 - e^{-t/\tau}). \tag{A.3}$$

It is evident from (A.3) that after some time the output will settle down at a steady state value (velocity) KA (i.e., rad/s). The settling time depends on the time constant of the system (τ). The smaller is τ the faster the system will reach its steady state target (final velocity) KA.

Now, let us consider that the command to the system is sent from a computer. A numerical value for input x is sent at discrete time intervals T from the computer to the physical system (i.e., motor amplifier). The number is converted to a voltage by a digital to analog (D/A) converter circuit. However, the computer sends the values only at T [s] intervals, and the value given to the physical device (i.e., amplifier) remains constant during the time interval T. The command generation is now completely in the discrete time domain; hence the input must be converted to the z domain. The z transform of step input $x = A$ is $Z[x(kT)] = X(z) = Z(A) = A/(1 - z^{-1})$. Although the amplifier–servo motor system receives the input commands in discrete time intervals T, it works in the continuous time domain since it is a physical system. To combine the discrete (digital computer) and continuous (i.e., servo motor) parts under one domain, we have to obtain a discrete equivalent transfer function of the physical system. The D/A converter circuit is the physical bridge between the two domains, and it has to be considered. It is most practical to model the D/A circuit by a sample and hold with a zero-order approximation, that is, $ZOH = (1 - e^{-sT})/s$. Since $z^{-1} = e^{-sT}$, a zero order hold equivalent of the continuous physical system is converted into the discrete time domain as follows:

$$G(z) = (1 - z^{-1})Z\frac{G(s)}{s} = (1 - z^{-1})\left[K\frac{1/\tau}{s(s + 1/\tau)}\right]$$

$$= (1 - z^{-1})\frac{Kz^{-1}(1 - e^{-T/\tau})}{(1 - z^{-1})(1 - e^{-T/\tau}z^{-1})} \tag{A.4}$$

$$= \frac{z^{-1}[K(1 - e^{-T/\tau})]}{1 - e^{-T/\tau}z^{-1}}.$$

Now that the system is completely transferred to z domain, we can analyze its response for a step input generated by the computer. The steady state response of the system for a step input with amplitude A can be found as follows:

$$y_{ss} = \lim_{z=1} X(z) \cdot G(z) = \lim_{z=1} \frac{A}{1 - z^{-1}} \cdot \frac{z^{-1}[K(1 - e^{-T/\tau})]}{1 - e^{-T/\tau}z^{-1}} = KA, \tag{A.5}$$

which is identical to the value found from (A.3) for the continuous system. We can look at the system response at discrete time intervals $k = 0, 1, 2, 3, \ldots, k$ that are equally spaced with T [s] periods, called the *sampling* or *control interval*. If the computer sends the step change in the following order: $x(0) = 0$, $x(1) = A$, $x(2) = A$, $x(3) = A, \ldots, x(k) = A$, then noting that $z^{-1}x(k) = x(k - 1)$ and $z^{-2}x(k) = x(k - 2)$, the system's response in discrete time intervals can be found from (A.5):

$$y(k) = G(z)x(k),$$

$$[(1 - e^{-T/\tau})z^{-1}]y(k) = z^{-1}[K(1 - e^{-T/\tau})]x(k), \tag{A.6}$$

$$y(k) = +e^{-T/\tau}y(k - 1) + [K(1 - e^{-T/\tau})]x(k - 1).$$

The reader can simulate both systems (Eqs. A.3 and A.7) by assigning numerical values to the system and comparing the results (e.g., $A = 10$, $K = 2$, $T = 1$, $\tau = 0.2$).

The handling of Laplace (continuous) and z domain transformations must be clearly understood before designing machine tool or machining process control algorithms.

In this appendix, besides introducing the derivation of the (inverse) Laplace and z transforms we will also present (inverse) Laplace and z transforms using the MATLAB$^{\copyright}$ Symbolic Math Toolbox [1].

A.2 BASIC DEFINITIONS

Let $x(t)$ be a continuous function of time t with sampled values represented by the sequence

$$\{x(kT)\} = \{x(0), x(T), x(2T), \ldots\}$$

or (A.7)

$$\{x(k)\} = \{x(0), x(1), x(2), \ldots\},$$

where T is the sampling period, which is sometimes not explicitly indicated. In Eq. A.7, it is assumed that

$$x(t) = 0 \quad \text{for } t < 0 \quad \text{and} \quad x(kT) = 0 \quad \text{for } k < 0, \tag{A.8}$$

where k is a nonnegative integer $(k = 0, 1, 2, \ldots)$.

The Laplace transform of $x(t)$ is defined as

$$\mathcal{L}[x(t)] = X(s) = \int x(t)e^{-st}dt. \tag{A.9}$$

The z transform of $x(t)$ or of the corresponding sequence $\{x(kT)\}$ (see Eq. A.7) is defined as

$$Z[x(t)] = X\{x(kT)\} = \sum_{k=0}^{\infty} x(kT)z^{-k},$$

or dropping the sampling period T, we have

$$Z[x(t)] = X\{x(k)\} = \sum_{k=0}^{\infty} x(k)z^{-k} \tag{A.10}$$

and $x(kT) = Z^{-1}[X(z)]$ is the inverse z transform. Both Laplace and z transforms given by (A.9) and (A.10) are one-sided transforms (i.e., the integral and summation in (A.8) and (A.10) start from zero), which are justified following (A.8). These one-sided transforms are the ones that are widely used in controls.

Conceptually the z transform is easier to understand than the Laplace transform since the former involves summation while the latter involves integration. For example, consider the ramp function:

$$x(t) = \begin{cases} At & t \geq 0, \\ 0 & \text{otherwise.} \end{cases} \tag{A.11}$$

or the corresponding unit ramp sequence

$$x(t)|_{t=kT} = x(kT) = \begin{cases} AkT & k = 0, 1, 2, \ldots, \\ 0 & \text{otherwise.} \end{cases} \tag{A.12}$$

Following (A.9) and (A.11) and, for example, using integration by parts, the Laplace transform of $x(t)$ is obtained as

$$X(s) = \mathcal{L}[At] = \int_0^\infty Ate^{-st}dt = \frac{A}{s^2}. \tag{A.13}$$

However, the z transform of (A.12) is a straightforward summation:

$$X(z) = \sum_{k=0}^\infty AkT \cdot z^{-k} = AT(z^{-1} + 2z^{-2} + 3z^{-3} + \cdots), \tag{A.14}$$

which converges for $|z| > 1$. However, (A.14) is not in a very useful form because it is an open-ended series (i.e., the summation has infinitely many terms). Using the power series the z transform (A.14) can be written in closed form as

$$X(z) = AT\frac{z^{-1}}{(1 - z^{-1})^2} = AT\frac{z}{(z - 1)^2}. \tag{A.15}$$

Although, as was mentioned, the z transform is conceptually easier to understand than the Laplace transform, in textbooks on controls, the Laplace transform is considered before the z transform. The reason is that although in the analysis and design of continuous-time control systems the Laplace transform suffices, in the case of digital control systems both Laplace and z transforms are necessary. This is because the digital control systems have both continuous-time (e.g., plant) and discrete-time (e.g., controller) components as illustrated in the previous section.

As shown in (A.15), the z transform of a function or a sequence can be written both using the negative or positive powers of z. One word of caution is that when the negative powers of z are used, one has to be careful about the zeros at the origin of the z-plane [75]. To this end consider

$$X(z) = \frac{z}{z - 3} = \frac{1}{1 - 3z^{-1}}, \tag{A.16}$$

which, when written in terms of the positive powers of z, clearly has a pole at $z = 3$ and a zero at the origin. However, when $X(z)$ is written in terms of the negative powers of z, although the pole at $z = 3$ is clearly seen, the zero at the origin may not be that obvious. In this appendix the positive powers of z will be used to obtain the residues. However, whenever the z transform table has to be used the expressions obtained will be written in terms of the negative powers of z.

Given a transfer function in the s domain, the corresponding transfer function in the z domain can be obtained using several different methods such as: i) backward difference method; ii) forward difference method; iii) matched pole-zero mapping method; iv) bilinear transformation (or Tustin) method (with frequency pre-warping); v) impulse invariance method; and vi) step invariance

TABLE A.1. Table of Laplace and z Transforms [75]

$x(t)$	$X(s)$	$x(kT)$	$X(z)$
Impulse	—	$\delta(k)=\left\{\begin{smallmatrix}1 & k=0\\0 & k\neq1\end{smallmatrix}\right\}$	1
Delay	—	$\delta(n-k)=\left\{\begin{smallmatrix}1 & k=n\\0 & k\neq n\end{smallmatrix}\right\}$	z^{-k}
1	$\frac{1}{s}$	1	$\frac{1}{1-z^{-1}}$
e^{-at}	$\frac{1}{s+a}$	e^{-akT}	$\frac{1}{1-e^{-aT}z^{-1}}$
t	$\frac{1}{s^2}$	kT	$\frac{Tz^{-1}}{(1-z^{-1})^2}$
t^2	$\frac{2}{s^3}$	$(kT)^2$	$\frac{T^2z^{-1}(1+z^{-1})}{(1-z^{-1})^3}$
t^3	$\frac{6}{s^4}$	$(kT)^3$	$\frac{T^3z^{-1}(1+4z^{-1}+z^{-2})}{(1-z^{-1})^4}$
$1-e^{-at}$	$\frac{a}{s(s+a)}$	$1-e^{-akT}$	$\frac{z^{-1}(1-e^{-aT})}{(1-z^{-1})(1-e^{-aT}z^{-1})}$
$e^{-at}-e^{-bt}$	$\frac{b-a}{(s+a)(s+b)}$	$e^{-akT}-e^{-bkT}$	$\frac{z^{-1}(e^{-aT}-e^{-bT})}{(1-e^{-aT}z^{-1})(1-e^{-bT}z^{-1})}$
te^{-at}	$\frac{1}{(s+a)^2}$	kTe^{-akT}	$\frac{Te^{-aT}z^{-1}}{(1-e^{-aT}z^{-1})^2}$
$(1-at)e^{-at}$	$\frac{s}{(s+a)^2}$	$(1-akT)e^{-akT}$	$\frac{1-(1+aT)e^{-aT}z^{-1}}{(1-e^{-aT}z^{-1})^2}$
t^2e^{-at}	$\frac{2}{(s+a)^3}$	$(kT)^2e^{-akT}$	$\frac{T^2e^{-aT}(1+e^{-aT}z^{-1})z^{-1}}{(1-e^{-aT}z^{-1})^3}$
$\sin\omega t$	$\frac{\omega}{s^2+\omega^2}$	$\sin\omega kT$	$\frac{z^{-1}\sin\omega T}{1-2z^{-1}\cos\omega T+z^{-2}}$
$\cos\omega t$	$\frac{s}{s^2+\omega^2}$	$\cos\omega kT$	$\frac{1-z^{-1}\cos\omega T}{1-2z^{-1}\cos\omega T+z^{-2}}$
$e^{-at}\sin\omega t$	$\frac{\omega}{(s+a)^2+\omega^2}$	$e^{-akT}\sin\omega kT$	$\frac{e^{-aT}z^{-1}\sin\omega T}{1-2e^{-aT}z^{-1}\cos\omega T+e^{-2aT}z^{-2}}$
$e^{-at}\cos\omega t$	$\frac{s+a}{(s+a)^2+\omega^2}$	$e^{-akT}\cos\omega kT$	$\frac{1-e^{-aT}z^{-1}\cos\omega T}{1-2e^{-aT}z^{-1}\cos\omega T+e^{-2aT}z^{-2}}$
		a^k	$\frac{1}{1-az^{-1}}$
		$a^{k-1},\ \ k=1,2,3\ldots$	$\frac{z^{-1}}{1-az^{-1}}$
		ka^{k-1}	$\frac{z^{-1}}{(1-az^{-1})^2}$
		$a^k\cos k\pi$	$\frac{1}{1+az^{-1}}$

method [75]. Among these methods the impulse invariance uses the z transform directly.

Note that the examples given by Eqs. (A.11) and (A.12) are elementary cases and, therefore, rather than using the transformation formulas as in (A.13) and (A.14), we could directly use Table A.1 to obtain the Laplace and z transforms. To simplify the problem in nonelementary cases, it is perhaps easier to use Laplace and z transform theorems given in Tables A.2 and A.3 whenever possible. In fact in Eqs. A.13 and A.15 we already made use of the first property of Laplace and z transform theorems.

TABLE A.2 Properties of Laplace Transforms [75]

$\mathcal{L}[Af(t)] = AF(s)$

$\mathcal{L}[f_1(t) \pm f_2(t)] = F_1(s) \pm F_2(s)$

$\mathcal{L} \pm \left[\frac{d}{dt} f(t)\right] = sF(s) - f(0\pm)$

$\mathcal{L} \pm \left[\frac{d^2}{dt^2} f(t)\right] = s^2 F(s) - sf(0\pm) - \dot{f}(0\pm)$

$\mathcal{L}[e^{-at} f(t)] = F(s+a)$

$\mathcal{L}[f(t-a)1(t-a)] = e^{-as} F(s)$

$\mathcal{L}[tf(t)] = -\frac{dF(s)}{ds}$

$\mathcal{L}\left[\frac{1}{t} f(t)\right] = \int_s^\infty F(s) ds$

Next we will consider another example by which we will demonstrate using the table of Laplace transform properties (Table A.2):

$$\mathcal{L}\{Ae^{-at} \cos \omega t\} = A \cdot \mathcal{L}\{e^{-at} \cos \omega t\} \tag{A.17}$$

where A, a, and ω are real. From the Laplace transform table (A.1) we have

$$\mathcal{L}\{\cos \omega t\} = \frac{s}{s^2 + \omega^2}, \tag{A.18}$$

and using the Laplace transform property of real translation in the s domain (see Table A.2) we obtain

$$\mathcal{L}\{Ae^{-at} \cos \omega t\} = A\frac{s+\alpha}{(s+\alpha)^2 + \omega^2}. \tag{A.19}$$

Note that the function considered "$e^{-at} \cos \omega t$" represents a damped sinusoidal signal and, owing to its importance in engineering systems, it is given in the Laplace transform Table A.1 as well.

In the rest of this appendix we will consider inverse Laplace and z transforms. In the case of inverse Laplace transforms attention will be focused on the partial fraction expansion method. In the case of the inverse z transform, besides the partial fraction expansion method, we will also consider the residue method. The residue method, which is also known as the inversion residue method as well as the inversion formula method, leads to the inverse z transforms directly without resorting to the z transform table.

TABLE A.3. Properties of z Transforms [75]

$x(k)$ or $x(k)$	$Z[x(t)]$ or $Z[x(k)]$
$ax(t)$	$aX(z)$
$ax_1(t) + bx_2(t)$	$aX_1(z) + bX_2(z)$
$x(t+T)$ or $x(k+1)$	$zX(z)z - x(0)$
$x(t+2T)$	$z^2 X(z) - z^2 x(0) - zx(T)$
$x(t+kT)$	$z^k X(z) - z^k x(0) - z^{k-1} x(T) - \cdots - zx(kT - T)$
$x(t-kT)$ or $x(n-k)$	$z^{-k} X(z)$
$tx(t)$	$-Tz\frac{d}{dz} X(z)$
$e^{-at} x(t)$	$X(ze^{aT})$
$\nabla x(k) = x(k) - x(k-1)$	$(1 - z^{-1})X(z)$
$\sum_{k=0}^n x(k)$	$\frac{1}{1-z^{-1}} X(z)$

A.3 PARTIAL FRACTION EXPANSION METHOD

The partial fraction expansion method is a mathematical tool similar to the continued fraction method. The partial fraction expansion method is also utilized to obtain inverse Laplace, z, and Fourier transforms among which we will consider the first two.

In the case of inverse transforms, the partial fraction expansion method enables us to write a given a rational function in s or z in such a way that we can use the Laplace or z transform tables directly or after some manipulations.

Consider the transfer function

$$X(p) = \frac{N(p)}{(p - p_1) \cdots (p - p_r) \cdots (p - p_n)}, \tag{A.20}$$

which is assumed to have distinct poles and no pole-zero cancellation. In (A.20) the degree of the numerator polynomial $N(p)$ is less than the degree of the denominator polynomial, which is n. In other words $X(p)$ is a strictly proper rational function.

Note that in the case of the Laplace transform we will set $p = s$ and in the case of the z transform we will set $p = z$. In the rest of this appendix we will consider the negative powers of z only when we use the z transform table.

If the degree of the numerator polynomial $N(p)$ of (A.20) is equal to or greater than n, long division is used yielding

$$X(p) = F(p) + \breve{X}(p), \tag{A.21}$$

where $F(p)$ is a polynomial of positive powers of p and is a strictly proper rational function. If $X(p)$ represents a transfer function of an engineering system, following the causality condition, $F(p)$ will be a constant.

In the case of distinct poles, partial fraction expansion of (A.20) is written as

$$X(p) = \frac{A_1}{p - p_1} + \cdots + \frac{A_r}{p - p_r} + \cdots + \frac{A_n}{p - p_n}, \tag{A.22}$$

where $A_i(i = 1, \ldots, n)$ is the residue associated with pole p_i. Noting that all poles in (A.22) are assumed to be distinct, the residues are obtained according to

$$A_i = [(p - p_i)X(p)]|_{p=p_i}, \tag{A.23}$$

where (A.23) is obtained by multiplying (A.22) by $p - p_i$ and evaluating at $p = p_i$. Note that (A.22) and (A.23) are valid when poles are real or complex conjugate $(i = 1, 2, \ldots, n)$. In the later case, for example, if $p_2 = p_1^*$ then $A_2 = A_1^*$ where the asterisk indicates complex conjugate.

In the case of multiple poles

$$X(p) = \frac{A_1}{p - p_1} + \frac{A_{21}}{p - p_2} + \frac{A_{22}}{(p - p_2)^2} + \cdots + \frac{A_{2j}}{(p - p_2)^j} + \cdots + \frac{A_{2r}}{(p - p_2)^r}, \tag{A.24}$$

where the multiplicity of pole p_2 is assumed to be r; the residues associated with the multiple pole are derived from

$$A_{2j} = \frac{1}{(r-j)!} \left\{ \frac{d^{r-j}}{dp^{r-j}}[(p-p_2)^r X(p)] \right\} \Bigg|_{p=p_r} \quad (j = 1, \ldots, r). \tag{A.25}$$

Note that the residues can alternatively be obtained by solving a set of algebraic equations, as will be demonstrated in Example 1.

The following examples demonstrate obtaining the residues using the partial fraction method.

Example 1. To demonstrate the partial fraction expansion for single poles, consider the following rational function with distinct poles:

$$X(p) = \frac{1}{(p+2)(p+4)} = \frac{A_1}{p+2} + \frac{A_2}{p+4}. \tag{A.26}$$

Following (A.23) and (A.26) can be written as

$$X(p) = \frac{0.5}{p+2} + \frac{-0.5}{p+4}. \tag{A.27}$$

Note that the residues $A_1 = 0.5$ and $A_2 = -0.5$ alternatively could also be obtained from a set of linear equations:

$$\begin{bmatrix} 1 & 1 \\ 4 & 2 \end{bmatrix} \cdot \begin{bmatrix} A_1 \\ A_2 \end{bmatrix} = \begin{bmatrix} 0 \\ 1 \end{bmatrix}, \tag{A.28}$$

which obtained from

$$\frac{A_1(p+4) + A_2(p+2)}{(p+2)(p+4)} = \frac{1}{(p+2)(p+4)}, \tag{A.29}$$

where we equate the equal powers of p in the numerator.

Example 2. Next consider the partial fraction expansion in the case of multiple poles:

$$X(p) = \frac{p+0.4}{(p+0.1)^2(p+0.2)}, \tag{A.30}$$

which, using (A.25), yields

$$X(p) = \frac{-20}{p-0.1} + \frac{3}{(p-0.1)^2} + \frac{20}{p-0.2}, \tag{A.31}$$

where each term of (A.31) appears in the Laplace transform table (see Example 5).

Example 3. In the case of inverse z transforms, the poles at the origin have to be handled carefully. Therefore, in this example we chose a transfer function with a double pole at the origin:

$$X(p) = \frac{1}{(p - 0.8)^2 p^2}. \tag{A.32}$$

This transfer function will be considered in Examples 8 and 10. Using the residue formulas (A.23) and (A.25) the partial fraction expansion of (A.32) can be written as

$$X(p) = \frac{3.9}{p - 0.8} + \frac{1.56}{(p - 0.8)^2} + \frac{3.9}{p} + \frac{1.56}{p^2}. \tag{A.33}$$

A.4   PARTIAL FRACTION EXPANSION METHOD TO DETERMINE INVERSE LAPLACE AND z TRANSFORMS

In this section we will employ the partial fraction expansion method to obtain the inverse Laplace and z transforms. Note that in the case of the z transform, if the transfer function $X(z)$ is written using the positive powers of z and has at least one zero at the origin, than instead of $X(z)$ we can also consider $X(z)/z$ [75].

Example 4. In order to demonstrate the inverse Laplace transform in the case of single poles, consider the transfer function (A.26) in the s domain by substituting $p = s$:

$$X(s) = \frac{1}{(s + 2)(s + 4)}. \tag{A.34}$$

Following (A.27), we obtain $x(t)$ using the Laplace transform table as

$$x(t) = \mathcal{L}^{-1}[X(s)] = 0.5\mathcal{L}^{-1}\frac{1}{s + 2} - 0.5\mathcal{L}^{-1}\frac{1}{s + 4} = 0.5e^{-2t}u(t) - e^{-4t}u(t), \tag{A.35}$$

where $u(t)$ is the unit step.

Example 5. Next we consider the inverse Laplace transform of transfer functions (A.30) with multiple poles:

$$X(s) = \frac{s + 0.4}{(s + 0.1)^2(s + 0.2)}. \tag{A.36}$$

Following (A.31) and using the Laplace transform table, we have

$$x(t) = \mathcal{L}^{-1}[X(s)] = -20\mathcal{L}^{-1}\frac{1}{s + 0.1} + 3\mathcal{L}^{-1}\frac{1}{(s + 0.1)^2} + 20\mathcal{L}^{-1}\frac{1}{s + 0.2}$$
$$= (-20e^{-0.1t} + 3te^{-0.1t} + 20e^{-0.2t})u(t). \tag{A.37}$$

Example 6. To demonstrate the inverse z transform we start by considering the transfer functions with single poles as in the case of the inverse Laplace transform (see Example 4). Letting $p = z$ in (A.26) we get

$$X(z) = \frac{1}{(z + 2)(z + 4)} = \frac{1}{z^2 + 6z + 8}. \tag{A.38}$$

Equation (A.27) can be written in Z^{-1} domain by multiplying the numerator and denominator by z^{-2}:

$$X(z) = \frac{z^{-2}}{1 + 6z^{-1} + 8z^{-2}}. \tag{A.39}$$

For discrete input $u(k)$, the output $x(k)$ can be evaluated at each step k as a recursive function of past input and output values, that is,

$$x(k) = u(k-2) - 6x(k-1) - 8x(k-2). \tag{A.40}$$

However, the contributions of each root-pole on both transfer function as well as on the value of each response for any given input state can be explained better by looking at the partial fraction of the transfer function. Equation (A.27) has the following partial fraction expansion:

$$X(z) = \frac{0.5}{z+2} + \frac{-0.5}{z+4}. \tag{A.41}$$

Note that if the transfer function is given in terms of the negative powers of z, we will write the transfer function in terms of the positive powers of z before deriving the partial fractions.

From (A.41) we have

$$x(k) = Z^{-1}[X(z)] = 0.5Z^{-1}\frac{1}{z+2} - 0.5Z^{-1}\frac{1}{z+4}$$

$$= 0.5Z^{-1}\frac{z^{-1}}{1+2z^{-1}} - 0.5Z^{-1}\frac{z^{-1}}{1+4z^{-1}}, \tag{A.42}$$

which, using the tables and simplifying the results obtained, yields

$$x(k) = (-1)^{k-1}\{2^{k-2} - 2^{2k-3}\}u(k-1). \tag{A.43}$$

Here $u(k-1)$ is the one-sample delayed unit step sequence.

Note that even though partial fractions are obtained using positive powers of z, the inverse z transform is obtained from the table that is given in terms of the negative powers of z.

Example 7. Consider the rational function (A.30) with multiple poles, to demonstrate the inverse z transform. Substituting $p = z$ we get

$$X(z) = \frac{z+0.4}{(z+0.1)^2(z+0.2)} \tag{A.44}$$

and from (A.31) we have

$$x(k) = Z^{-1}[X(z)] = -20Z^{-1}\frac{1}{z+0.1} + 3Z^{-1}\frac{1}{(z+0.1)^2} + 20Z^{-1}\frac{1}{z+0.2}$$

$$= -20Z^{-1}\frac{z^{-1}}{1+0.1z^{-1}} + 3Z^{-1}\frac{z^{-2}}{(1+0.1z^{-1})^2} + 20Z^{-1}\frac{z^{-1}}{1+0.2z^{-1}}. \tag{A.45}$$

Considering the second term on the right-hand side of (A.45) and using the z

transform and the shifting property we have

$$Z^{-1}\left[z^{-1}\left(\frac{z^{-1}}{(1+0.1z^{-1})^2}\right)\right] = (k-1)(-0.1)^{k-2}u(k-1), \tag{A.46}$$

from which using (A.45) we obtain

$$x(k) = \{1 + 2^k - 3k\}(-1)^{k-1}(0.1)^{k-2}u(k-1). \tag{A.47}$$

Example 8. As a last example to demonstrate the use of the partial fraction expansion in obtaining the inverse z transform of transfer functions with multiple poles at the origin, consider the rational function (A.32) with $p = z$:

$$X(z) = \frac{1}{(z-0.8)^2 z^2}. \tag{A.48}$$

Following (A.33), we have

$$X(z) = \frac{3.9}{z-0.8} + \frac{1.56}{(z-0.8)^2} + \frac{3.9}{z} + \frac{1.56}{z^2}, \tag{A.49}$$

$$x(k) = Z^{-1}[X(z)] = 3.9Z^{-1}\frac{z^{-1}}{1-0.8z^{-1}}$$

$$+ 1.56Z^{-1}\frac{z^{-2}}{(1-0.8z^{-1})^2} + 3.9Z^{-1}[z^{-1}] + 1.56Z^{-1}[z^{-2}], \tag{A.50}$$

and using

$$Z^{-1}\left[z^{-1}\frac{z^{-1}}{(1-0.8z^{-1})^2}\right] = (k-1)(0.8)^{k-2}u(k-1),$$

$$Z^{-1}\left[\frac{z^{-1}}{1-0.8z^{-1}}\right] = (0.8)^{k-1}u(k-1),$$

$$Z^{-1}[z^{-1}] = \delta(k-1) = \left\{\begin{matrix} 1 & k=1, \\ 0 & k\neq 1, \end{matrix}\right\}$$

$$Z^{-1}[z^{-2}] = \delta(k-2) = \left\{\begin{matrix} 1 & k=2, \\ 0 & k\neq 2, \end{matrix}\right\}$$

we obtain

$$x(k) = (k-3)(0.8)^{k-4}u(k-3). \tag{A.51}$$

The transfer function (A.48) will be reconsidered later on in Example 10 using the residue method, which is presented next.

A.5 THE RESIDUE METHOD TO DETERMINE INVERSE z TRANSFORM

The residue method is also known as the *inversion formula method* and the *inversion integral method*. Given a rational function $X(z)$, the residue method directly leads to inverse z transform $x(k)$ without using the z transform table.

The inverse z transform $x(k)$ is obtained using the formula

$$x(k) = \sum \{\text{residues of } z^{k-1}X(z)\}, \tag{A.52}$$

where the summation is over all poles of $z^{k-1}X(z)$.

Consider a typical transfer function with single and multiple poles:

$$X(z) = \frac{N(z)}{(z - p_1)(z - p_2)^r}. \tag{A.53}$$

To use the residue method we need two residue formulas, as in the case of partial fraction expansion, one for the single pole p_1:

$$\{\text{residues of } z^{k-1}X(z)\}|_{z=p_1} = \{(z - p_1)z^{k-1}X(z)\}|_{z=p_1} \tag{A.54}$$

and the other for the multiple poles at p_2:

$$\{\text{residues of } z^{k-1}X(z)\}_{z=p_2} = \left\{ \frac{1}{(r-1)!} \frac{d^{r-1}}{dz^{r-1}} [(z - p_2)^r z^{k-1} X(z)] \right\} \bigg|_{z=p_2}. \tag{A.55}$$

Considering the multiple poles, note that for the residue formula (A.55) does not depend on an index j as in the partial fraction expansion case. Also note that formula (A.55) involves $z^{k-1}X(z)$ not $X(z)$. The z^{k-1} term results in a pole at the origin for $k = 0$ if $X(z)$ does not have a zero at the origin. Furthermore, the z^{k-1} term will change the number of poles at the origin if $X(z)$ has poles at the origin. Therefore, particular attention has to be paid to the pole(s) at the origin.

Example 9. Considering the transfer function given in Example 6:

$$X(z) = \frac{1}{(z + 2)(z + 4)}. \tag{A.56}$$

We will obtain the inverse z transform using the residue method. First we note that the transfer function (A.56) has single poles and has no zero or pole at the origin of the z plane. Following (A.52) we form

$$k^{k-1}X(z) = \frac{z^{k-1}}{(z + 2)(z + 4)}, \tag{A.57}$$

which for $k = 0$ becomes

$$z^{k-1}X(z)|_{k=0} = \frac{1}{z(z + 2)(z + 4)}. \tag{A.58}$$

Note that, compared to (A.56) and (A.58) has an additional pole at the origin. Therefore, the cases $k = 0$ and $k \geq 1$ must be treated separately. Considering first $k = 0$, we have

$$x(0) = \left\{ z \frac{1}{z(z + 2)(z + 4)} \right\} \bigg|_{z=0} + \left\{ (z + 2) \frac{1}{z(z + 2)(z + 4)} \right\} \bigg|_{z=-2}$$
$$+ \left\{ (z + 4) \frac{1}{z(z + 2)(z + 4)} \right\} \bigg|_{z=-4} = 0. \tag{A.59}$$

For $k \geq 1$,

$$x(k) = \left\{ (z+2)\frac{z^{k-1}}{(z+2)(z+4)} \right\}\Big|_{z=-2} + \left\{ (z+4)\frac{z^{k-1}}{(z+2)(z+4)} \right\}\Big|_{z=-4}$$

$$= (-1)^{k-1}\{2^{k-2} - 2^{2k-3}\}u(k-1). \tag{A.60}$$

In this case (A.60) holds for $k > 0$ and agrees with the result obtained using partial fraction expansion (see Eq. A.43).

Example 10. Consider the transfer function considered in Example 8:

$$X(z) = \frac{1}{(z-0.8)^2 z^2}, \tag{A.61}$$

which has multiple poles at the origin. Here we will obtain the inverse z transform of (A.61) using the residue method. Following (A.52) we consider

$$z^{k-1}X(z) = \frac{z^{k-1}}{(z-0.8)^2 z^2}, \tag{A.62}$$

which takes the form

$$z^{k-1}X(z) = \frac{1}{(z-0.8)^2 z^3} \quad \text{(for } k = 0),$$

$$X(z) = \frac{1}{(z-0.8)^2 z^2} \quad \text{(for } k = 1),$$

$$zX(z) = \frac{1}{(z-0.8)^2 z} \quad \text{(for } k = 2),$$

$$z^{k-1}X(z) = \frac{z^{k-3}}{(z-0.8)^2} \quad \text{(for } k \geq 3).$$

Hence, the residues at $k = 0, 1, 2$ and $k \geq 3$ are derived as

$$x(0) = \left\{ \frac{d}{dz}z^{-3} \right\}\Big|_{z=0.8} + \left\{ \frac{1}{2}\frac{d^2}{dz^2}(z-0.8)^{-2} \right\}\Big|_{z=0} = 0,$$

$$x(1) = \left\{ \frac{d}{dz}z^{-2} \right\}\Big|_{z=0.8} + \left\{ \frac{d}{dz}(z-0.8)^{-2} \right\}\Big|_{z=0} = 0,$$

$$x(2) = \left\{ \frac{d}{dz}z^{-1} \right\}\Big|_{z=0.8} + \left\{ (z-0.8)^{-2} \right\}\Big|_{z=0} = 0,$$

$$x(k) = (k-3)(0.8)^{k-4} \quad \text{for } k \geq 3.$$

This leads to

$$x(k) = (k-3)(0.8)^{k-4}u(k-3), \tag{A.63}$$

which agrees with (A.51) of Example 8.

A.6 DERIVATION OF THE INVERSE LAPLACE AND *z* TRANSFORMS USING MATLAB© SYMBOLIC MATH TOOLBOX

In this section we will reconsider the above examples and solve them using MATLAB© Symbolic Math Toolbox [1]. In the following, we will adopt the notation used in the above sections when using MATLAB©. To this end, the Laplace, inverse Laplace, *z*, and inverse *z* transforms will be obtained using the functions *laplace(x, t, s), ilaplace(X, s, t), ztrans(x, n, z),* and *iztrans(X, z, n),* respectively.

To simplify the results obtained, MATLAB© Symbolic Math Toolbox has some functions such as "simplify" (simplifies) and "simple" (searches for the shortest expression), where the latter is more powerful than the former. Another MATLAB© function *"pretty"* puts the expression in a more familiar form.

Reconsider Example 4

EDU≫ syms s t   % define symbolic variables, s for Laplace and t for time

EDU≫ X4=
1/((s+2)*(s+4))   % enter the function in s domain
X4=1/(s+2)/(s+4)

% Note that if desired, the output X1 can be changed to a more familiar format by

EDU≫ pretty(X4)

$$\frac{1}{(s + 2)\,(s + 4)}$$

EDU≫ x4=ilaplace(X4,s,t)   % inverse Laplace transform
x4=
1/2*exp(-2*t)-1/2*exp(-4*t)

Reconsider Example 5

EDU≫ syms s t   % define symbolic variables
EDU≫ X5=(s+0.4)/((s+0.1)^2*(s+0.2)) % enter the function in s domain
X5=
(s+2/5)/(s+1/10)^2/(s+1/5)
EDU≫ x5=ilaplace(X5,s,t) % inverse Laplace transform
x5=
3*t*exp(-1/10*t)-20*exp(-1/10*t)+20*exp(-1/5*t)
EDU≫ pretty(x5)
3*t*exp(- 1/10*t) - 20*exp(- 1/10*t) + 20*exp(- 1/5*t)

Reconsider Example 6

EDU≫ syms n z
EDU≫ X6=1/((z+2)*(z+4))
X6=
1/(z+2)/(z+4)
EDU≫ x6=iztrans(X6,z,n)
x6=
1/8*Delta(n)-1/4*(-2)^n+1/8*(-4)^n
% where "Delta(n)" is the Kronecker delta

Reconsider Example 7

EDU≫ syms n z
EDU≫ X7=(z+0.4)/((z+0.1)^2*(z+0.2))
X7=
(z+2/5)/(z+1/10)^2/(z+1/5)
EDU≫ x7=iztrans(X7)
x7=
200*Delta(n)-100*(-1/10)^n+300*(-1/10)^n*n-100*(-1/5)^n
EDU≫ simplify(x7) % simplifying the output to a certain extent
ans =
200*Delta(n)+100*(-1)^(1+n)*(1/10)^n+300*(-1)^n*(1/10)^n*n+100*(-1)^
(1+n)*(1/5)^n

Reconsider Example 8 (or 10)

EDU≫ syms n z
EDU≫ X8=1/((z-0.8)^2*(z^2))
X8=
1/(z-4/5)^2/z^2
EDU≫ x8=iztrans(X8)
x8=
25/16*Delta(n-2)+125/32*Delta(n-1)+1875/256*Delta(n)-1875/256*(4/5)^
n+625/256*(4/5)^n*n
EDU≫ simplify(x8)
ans =
25/16*Delta(n-2)+125/32*Delta(n-1)+1875/256*Delta(n)-1875/256*4^
n*5^(-n)+625/256*4^n*5^(-n)*n

All the results obtained in this section agree with the results given in the former sections, although some of the results need some manipulations in order to compare them.

OFF-LINE AND ON-LINE PARAMETER ESTIMATION WITH LEAST SQUARES

B.1 OFF-LINE LEAST SQUARES ESTIMATION

The transfer function of systems can be identified either using impulse, step, and frequency response techniques in the continuous or discrete time domain or using regression techniques in the discrete time domain. When the process is to be controlled by a computer, it may be easier to identify the transfer function parameters of the system by collecting input and output data at uniform discrete time intervals that are equal to the computer control period.

Consider a general transfer function with d samples delay, with orders nb, and na in the numerator and denominator, respectively (such a system is called ARIMA(na, d, nb) in the literature):

$$G_o(z^{-1}) = \frac{z^{-d}(b_0 + b_1 z^{-1} + b_2 z^{-2} + \cdots + b_{nb} z^{-nb})}{1 + a_1 z^{-1} + a_2 z^{-2} + \cdots + a_{na} z^{-na}}. \tag{B.1}$$

The dead time can be easily observed by applying step, pulse, or harmonic input to the system and then measuring the time delay between the input and output on a digital scope. It is usually difficult to guess the order of a transfer function (i.e., d, na, nb) if we don't have any engineering feeling about the process. In such cases, one has to resort to stochastic order identification techniques, which are not covered here. In machine tool engineering however, the orders of drives are usually low, up to about second or third order if we neglect the dynamic characteristics of the structure. Furthermore, a good engineer has to have the basic understanding of the drive design and control process provided in Chapter Five. It is customary to start with a low order transfer function assumption and increase the order if the identification does not yield good output estimates with the measurements.

Let us assume a simple process with the following discrete transfer function:

$$G(z^{-1}) = \frac{y(k)}{u(k)} = \frac{z^{-2}(b_0 + b_1 z^{-1})}{1 + a_1 z^{-1}}, \tag{B.2}$$

which is a first-order process whose output (y) has a two sample period delay in responding to an input (u). By expanding the transfer function, we get

$$\begin{aligned}
(1 + a_1 z^{-1})y(k) &= [z^{-2}(b_0 + b_1 z^{-1})]u(k), \\
y(k) &= -a_1 y(k-1) + b_0 u(k-2) + b_1 u(k-3),
\end{aligned} \tag{B.3}$$

or

$$y(k) = \{-y(k-1) \quad u(k-2) \quad u(k-3)\}\{a_1 \quad b_0 \quad b_1\}^T = \phi(k)^T \theta, \tag{B.4}$$

where k is the sampling counter, $\phi(k)^T = \{-y(k-1) \quad u(k-2) \quad u(k-3)\}$ is the regression or measurement vector, and $\theta = \{a_1 \quad b_0 \quad b_1\}^T$ is the unknown parameter vector. The three unknown parameters are *estimated* from the N measurements, which are greater than the number of unknowns (i.e., $N \gg 3$). If $\hat{\theta} = \{\hat{a}_1 \quad \hat{b}_0 \quad \hat{b}_1\}$ are the estimated parameters, the error between the estimated and actual measurement at interval k is

$$\epsilon(k) = y(k) - \hat{y}(k) = y(k) - \phi(k)^T \hat{\theta}. \tag{B.5}$$

Assume that a pulse or random input containing a wide range of frequencies is given to the system, and N number of inputs and outputs are collected at fixed time intervals. The following set of expressions can be written for each measurement:

$$\begin{Bmatrix} \epsilon(3) \\ \epsilon(4) \\ \vdots \\ \epsilon(k) \\ \vdots \\ \epsilon(N) \end{Bmatrix} = \begin{Bmatrix} y(3) \\ y(4) \\ \vdots \\ y(k) \\ \vdots \\ y(N) \end{Bmatrix} - \begin{bmatrix} \{-y(2) & u(1) & u(0)\} \\ \{-y(3) & u(2) & u(1)\} \\ \vdots & \vdots & \vdots \\ \{-y(k-1) & u(k-2) & u(k-3)\} \\ \vdots & \vdots & \vdots \\ \{-y(N-1) & u(N-2) & u(N-3)\} \end{bmatrix} \begin{Bmatrix} a_1 \\ b_0 \\ b_1 \end{Bmatrix},$$

or, in matrix form,

$$Y = \Phi\theta, \tag{B.6}$$

where

$$Y = \{y(3) \quad y(4) \quad \cdots y(k) \quad \cdots y(N)\}^T,$$
$$\Phi = \{\phi(3) \quad \phi(4) \quad \cdots \phi(k) \quad \cdots \phi(N)\}^T.$$

The least squares identification method is based on minimizing the sum of the squares of errors evaluated from all measurements [90]. The sum of the squares of errors, or *cost function*, to be minimized is

$$J(\theta, N) = \sum_{k=3}^{N} \epsilon(k)^2 = (Y - \Phi\theta)^T(Y - \Phi\theta) = (Y^T - \theta^T\Phi^T)(Y - \Phi\theta). \tag{B.7}$$

By expanding the expression further, we have

$$J(\theta, N) = Y^TY - Y^T\Phi\theta - \theta^T\Phi^TY - \theta^T\Phi^T\Phi\theta.$$

The parameters are obtained by minimizing the error or cost function, that is,

$$\frac{\partial J(\theta, N)}{\partial \theta} = \begin{Bmatrix} \dfrac{\partial J(\theta, N)}{\partial \theta_1} \\ \dfrac{\partial J(\theta, N)}{\partial \theta_2} \end{Bmatrix} = -2\Phi^TY + 2\Phi^T\Phi\theta = 0,$$

which leads to the estimates of parameters from N measurements:

$$\theta = [\Phi^T \Phi]^{-1} \Phi^T Y. \tag{B.8}$$

The least squares technique presented here is a useful tool for identifying the parameters of a discrete transfer function with any order. The matrix multiplication and inversion can be evaluated by either writing a short computer code or by using popular software tools.

B.2 RECURSIVE PARAMETER ESTIMATION ALGORITHM

An on-line, recursive, regularized constant-trace algorithm is given below [90]:

$$\hat{\theta}(t) = \hat{\theta}(t-1) + a(t)k(t)(F_\mathrm{p}(t) - \phi^T(t)\hat{\theta}(t-1)), \tag{B.9}$$

$$k(t) = P(t-1)\phi(t)(1 + \phi^T(t)P(t-1)\phi(t))^{-1}, \tag{B.10}$$

$$\bar{P}(t) = \bar{P}(t-1) - a(t)k(t)\phi^T(t)P(t-1), \tag{B.11}$$

$$P(t) = c_1 \frac{\bar{P}(t)}{\mathrm{tr}(\bar{P}(t))} + c_2 I, \tag{B.12}$$

$$a(t) = \begin{cases} \bar{a} & \text{if } |F_\mathrm{p}(t) - \phi^T(t)\hat{\theta}(t-1)| > 2\delta, \\ 0 & \text{otherwise,} \end{cases} \tag{B.13}$$

where $\hat{\theta}(t) = [\hat{a}_1\ \hat{a}_2\ \hat{b}_0\ \hat{b}_1\ \hat{b}_2]^T$ is the estimated parameter vector, $k(t)$ is the estimation gain, $\phi(t) = [-F_\mathrm{p}(k-1)\ -F_\mathrm{p}(k-2)\ f_\mathrm{c}(k-1)\ f_\mathrm{c}(k-2)\ f_\mathrm{c}(k-3)]^T$ is the regression vector or observation vector, $\bar{P}(t)$ is the auxiliary covariance matrix, $\mathrm{tr}(\bar{P}(t)) = \sum_{i=1}^{5} \bar{P}_{ii}$ is the trace of the auxiliary covariance matrix, and $P(t)$ is the covariance matrix. Here $c_1 > 0$, $c_2 \geq 0$, and δ is an estimate of the magnitude of the tolerable fluctuation of the output of the process or noise.

BIBLIOGRAPHY

[1] *MATLAB*© *Reference Guide*. The Math Works, Inc., 1992.

[2] E. J. A. Armarego and R. H. Brown. *The Machining of Metals*. Prentice-Hall, 1969.

[3] M. C. Shaw. *Metal Cutting Principles*. Oxford University Press, 1984.

[4] P. L. B. Oxley. *The Mechanics of Machining*. Ellis Horwood Limited, 1989.

[5] M. E. Merchant. Mechanics of the Metal Cutting Process, ii. Plasticity Conditions in Orthogonal Cutting. *Journal of Applied Physics*, vol. 16:318–324, 1945.

[6] E. H. Lee and B. W. Shaffer. Theory of Plasticity Applied to the Problem of Machining. *Journal of Applied Mechanics*, vol. 18:405–413, 1951.

[7] W. B. Palmer and P. L. B. Oxley. Mechanics of Orthogonal Machining. *Proceedings Institution of Mechanical Engineers*, vol. 173(no. 24):623–654, 1959.

[8] G. Boothroyd. Temperatures in Orthogonal Metal Cutting. *Proceedings Institution of Mechanical Engineers*, vol. 177:789–802, 1963.

[9] A. O. Tay, M. G. Stevenson, G. De Vahl Davis, and P. L. B. Oxley. A Numerical Method for Calculating Temperature Distributions in Machining, from Force and Shear Angle Measurements. *International Journal of Machine Tool Design and Research*, vol. 16:335–349, 1976.

[10] N. N. Zorev. Interrelationship between Shear Processes Occurring along Tool Face and on Shear Plane in Metal Cutting. *Proceedings of International Research in Production Engineering Research Conference*, pages 42–49, Sept. 1963.

[11] D. A. Stephenson. Material Characterization for Metal-Cutting Force Modeling. *Journal of Engineering Materials and Technology*, vol. 111:210–219, Apr. 1989.

[12] J. Krystof. *Berichte uber Betriebswissenschaftliche Arbeiten, Bd., 12*. VDI Verlag, 1939.

[13] M. E. Merchant. Mechanics of the Cutting Process. *Journal of Applied Physics*, vol. 16:318–324, 1945.

[14] M. E. Merchant. Basic Mechanics of the Metal Cutting Process. *Trans. ASME Journal of Applied Mechanics*, pages A–168–175, 1944.

[15] G. V. Stabler. Fundamental Geometry of Cutting Tools. *Proceedings of the Institution of Mechanical Engineers*, pages 14–26, 1951.

[16] W. K. Luk. The Direction of Chip Flow in Oblique Cutting. *International Journal of Production Research*, vol. 10:67–76, 1972.

[17] G. C. I. Lin, P. Mathew, P. L. B. Oxley, and A. R. Watson. Predicting Cutting Force for Oblique Machining Conditions. *Proceedings of the Institution of Mechanical Engineers*, vol. 196:141–148, 1982.

[18] E. Budak, Y. Altintas, and E. J. A. Armarego. Prediction of Milling Force Coefficients from Orthogonal Cutting Data. *Trans. ASME Journal of Manufacturing Science and Engineering*, vol. 118:216–224, 1996.

[19] E. Shamoto and Y. Altintas. Prediction of Shear Angle in Oblique Cutting with Maximum Shear Stress and Minimum Energy Principle. *1997 ASME International Mechanical Engineering Congress and Exposition*, MED-Vol. 6-1:121–128, 1997.

[20] G. V. Stabler. The Chip Flow Law and Its Consequences. *Advances in Machine Tool Design and Research*, pages 243–251, 1964.

[21] E. J. A. Armarego and R. C. Whitfield. Computer Based Modelling of Popular Machining Operations for Force and Power Predictions. *Annals of the CIRP*, 34:65–69, 1985.

[22] E. J. A. Armarego and M. Uthaichaya. A Mechanics of Cutting Approach for Force Prediction in Turning Operations. *Journal of Engineering Production*, 1(1):1–18, 1977.

[23] E. J. A. Armarego. *Material Removal Processes – An Intermediate Course*. The University of Melbourne, 1993.

[24] I. S. Jawahir and C. A. van Luttervalt. Recent Developments in Chip Control Research and Applications. *Annals of the CIRP*, 42/2:49–54, 1993.

[25] W. Kluft, W. Konig, C. A. Lutterwelt, K. Nagayama, and A. J. Pekelharing. Present Knowledge of Chip Form. *Annals of the CIRP*, 28/2:441–455, 1979.

[26] M. E. Martellotti. An Analysis of the Milling Process. *Transactions of the ASME*, 63:677–700, 1941.

[27] M. E. Martellotti. An Analysis of the Milling Process. Part II: Down Milling. *Transactions of the ASME*, 67:233–251, 1945.

[28] H. J. Fu, R. E. DeVor, and S. G. Kapoor. A Mechanistic Model for the Prediction of the Force System in Face Milling Operations. *Trans. ASME Journal of Engineering for Industry*, 106:81–88, Feb. 1984.

[29] W. A. Kline, R. E. DeVor, and W. J. Zdeblick. A Mechanistic Model for the Force System in End Milling with Application to Machining Airframe Structures. In *North American Manufacturing Research Conference Proceedings*, page 297, Dearborn, MI, 1980. Society of Manufacturing Engineers. Vol. XVIII.

[30] Y. Altintas and P. Lee. A General Mechanics and Dynamics Model for Helical End Mills. *Annals of the CIRP*, 45(1):59–64, 1996.

[31] Y. Altintas and A. Spence. End Milling Force Algorithms for CAD Systems. *Annals of the CIRP*, 40(1):31–34, 1991.

[32] A. Spence and Y. Altintas. A Solid Modeller Based Milling Process Simulation and Planning System. *Trans. ASME Journal of Engineering for Industry*, 116(1):61–69, 1994.

[33] E. J. A. Armarego, D. Pramanik, A. J. R. Smith, and Whitfield R. C. Forces and Power in Drilling – Computer Aided Predictions. *Journal of Production Engineering*, 6:149–174, 1983.

[34] D. J. Galloway. Some Experiments on the Influence of Various Factors on Drill Performance. *Transactions of the ASME*, 79:139, 1957.

[35] L. H. Chen and S. M. Wu. Further Investigation of Multifaceted Drills. *Trans. ASME Journal of Engineering for Industry*, 106:313–324, 1984.

[36] D. M. Ricon and A. G. Ulsoy. Effects of Drill Vibrations on Cutting Forces and Torque. *Annals of the CIRP*, 43(1):59–62, 1994.

[37] S. Kaldor and E. Lenz. Drill Point Geometry and Optimization. *Trans. ASME Journal of Engineering for Industry*, 104:84–90, 1982.

[38] W. Konig and K. Essel. New Tool Materials – Wear Mechanism and Application. *Annals of the CIRP*, 24/1:1–5, 1975.

[39] E. M. Trent. *Metal Cutting*. Butterworths, 1977.

[40] H. Ren. Mechanics of Machining with Chamfered Tools. Master's thesis, The University of British Columbia, 1998.

[41] F. W. Taylor. On the Art of Cutting Metals. *Transactions of the ASME*, 28, 1907.

[42] N. N. Zorev. *Metal Cutting Mechanics*. Pergamon. 1966.

[43] T. N. Loladze. Nature of Brittle Failure of Cutting Tool. *CRIP*, 24/1:13–16, 1975.

[44] G. Boothroyd. *Fundamentals of Machining and Machine Tools*. McGraw-Hill, 1975.

[45] F. Koenigsberger and J. Tlusty. *Machine Tool Structures – Vol. I: Stability Against Chatter*. Pergamon Press, 1967.

[46] S. Kalpakjian. *Manufacturing Engineering and Technology*. Addison-Wesley, 1995.

[47] S. Smith and J. Tlusty. An Overview of Modelling and Simulation of the Milling Process. *Trans. ASME Journal of Engineering for Industry*, 113(2):169–175, 1991.

[48] W. A. Kline, R. E. DeVor, and I. A. Shareef. The Prediction of Surface Accuracy in End Milling. *Trans. ASME Journal of Engineering for Industry*, 104:272–278, 1982.

[49] E. Budak and Y. Altintas. Flexible Milling Force Model for Improved Surface Error Predictions. In *Proceedings of the 1992 Engineering System Design and Analysis, Istanbul, Turkey*, pages 89–94. ASME, 1992. PD-Vol. 47-1.

[50] J. W. Sutherland and R. E. DeVor. An Improved Method for Cutting Force and Surface Error Prediction in Flexible End Milling Systems. *Trans. ASME Journal of Engineering for Industry*, 108:269–279, 1986.

[51] E. Budak and Y. Altintas. Peripheral Milling Conditions for Improved Dimensional Accuracy. *International Journal of Machine Tool Design and Research*, 34(7):907–918, 1994.

[52] R. I. King. *Handbook of High Speed Machining Technology*. Chapman and Hall, 1985.

[53] J. Tlusty and T. Moriwaki. Experimental and Computational Identification of Dynamic Structural Methods. *Annals of the CIRP*, 25:497–503, 1976.

[54] S. S. Rao. *Mechanical Vibrations*. Addison-Wesley, 1990.

[55] D. J. Ewins. *Modal Testing Theory and Practice*. Research Studies Press, 1984.

[56] R. J. Allemang and D. L. Brown. Multiple Input Experimental Modal Analysis – A Survey. *International Journal of Analytical and Experimental Modal Analysis*, 44:37–44, 1986.

[57] J. Tlusty and M. Polacek. The Stability of Machine Tools Against Self Excited Vibrations in Machining. *International Research in Production Engineering, ASME*, pages 465–474, 1963.

[58] S. A. Tobias and W. Fishwick. A Theory of Regenerative Chatter. *The Engineer – London*, 1958.

[59] S. A. Tobias. *Machine Tool Vibration*. Blackie, 1965.

[60] H. E. Merrit. Theory of Self-Excited Machine Tool Chatter. *Trans. ASME Journal of Engineering for Industry*, 87:447–454, 1965.

[61] J. Tlusty and F. Ismail. Basic Nonlinearity in Machining Chatter. *Annals of the CIRP*, 30:21–25, 1981.

[62] Y. Altintas and E. Budak. Analytical Prediction of Stability Lobes in Milling. *Annals of the CIRP*, 44(1):357–362, 1995.

[63] E. Budak and Y. Altintas. Analytical Prediction of Chatter Stability in Milling. Part i: General Formulation, Part ii: Application of the General Formulation to Common Milling Systems. *Trans. ASME Journal of Dynamic Systems, Measurement and Control*, 120:22–36, 1998.

[64] R. E. Hohn, R. Sridhar, and G. W. Long. A Stability Algorithm for a Special Case of the Milling Process. *Trans. ASME Journal of Engineering for Industry*, pages 325–329, May 1968.

[65] I. Minis, T. Yanushevsky, R. Tembo, and R. Hocken. Analysis of Linear and Nonlinear Chatter in Milling. *Annals of the CIRP*, 39:459–462, 1990.

[66] Y. Koren. *Computer Control of Manufacturing Systems*. McGraw Hill, 1983.

[67] I. Zeid. *CAD/CAM, Theory and Practice*. McGraw Hill, 1991.

[68] B. R. Dewey. *Computer Graphics for Engineers*. Harper & Row, 1988.

[69] CAMI. *Encyclopedia of the APT Programming Language*. Computer Aided Manufacturing International, Arlington, Texas, 1973.

[70] Y. Altintas. *Auto-Generation of a Finite Element Model from a Sculptured Surface Topological Model*. M.Sc. thesis, The University of New Brunswick, 1980.

[71] G. Stute. *Electrical Feed Drives for Machine Tools*. Edited by H. Gross, Wiley, 1983.

[72] THK. *THK LM System Ball Screws, Catalog No. 75-IBE-2*. T H K Co., Japan, 1996.

[73] Y. Altintas. Prediction of Cutting Forces and Tool Breakage in Milling from Feed Drive Current Measurements. *Trans. ASME Journal of Engineering for Industry*, vol. 114:386–392, 1992.

[74] K. Ogata. *Discrete Time Control Systems*. Prentice Hall, 1987.

[75] K. Ogata. *Modern Control Engineering*. Prentice Hall, 1970.

[76] Y. Altintas and N. A. Erol. Open Architecture Modular Tool Kit for Motion and Machining Process Control. *Annals of the CIRP*, 47(1):295–300, 1998.

[77] Y. Altintas and J. Peng. Design and Analysis of a Modular CNC System. *Computers in Industry*, 13:305–316, 1990.

[78] J. G. Bollinger and N. A. Duffie. *Computer Control of Machines and Processes*. Addison-Wesley, 1988.

[79] W. H. Press, B. P. Flannery, S. A. Tenkolsky, and W. T. Vetterling. *Numerical Recipes in C*. Cambridge University Press, 1988.

[80] M. Weck. *Handbook of Machine Tools – Automation of Controls*, volume 3. Wiley, 1984.

[81] Y. Altintas, N. Newell, and M. Ito. Modular CNC Design for Intelligent Machining. Part i: Design of a Hierarchical Motion Control Module for CNC Machine Tools. *ASME Journal of Manufacturing Science and Engineering*, 118:506–513, 1996.

[82] L. N. Srinivasan and Q. J. Ge. Parametric Continuous and Smooth Motion Interpolation. *Journal of Mechanical Design*, 118(4):494–498, 1996.

[83] F. C. Wang, S. Schofield, and P. Wright. Open Architecture Controllers for Machine Tools, Part 2: A Real Time Quintic Spline Interpolator. *Trans. ASME Journal of Manufacturing Science and Engineering*, 120(2):425–432, 1998.

[84] Y. Altintas and A. J. Lane. Design of an Electro-Hydraulic CNC Press Brake. *International Journal of Machine Tool Design and Research*, 37:45–59, 1997.

[85] H. E. Merrit. *Hydraulic Control Systems*. Wiley, 1967.

[86] J. K. Astrom and B. Wittenmark. *Computer Controlled Systems*. Prentice-Hall, 1984.

[87] Y. Altintas and W. K. Munasinghe. A Hierarchical Open Architecture CNC System for Machine Tools. *Annals of the CIRP*, 43(1):349–354, 1994.

[88] Y. Altintas and K. Munasinghe. Modular CNC System Design for Intelligent Machining, Part ii: Modular Integration of Sensor Based Milling Process Monitoring and Control Tasks. *Trans. ASME Journal of Manufacturing Science and Engineering*, 118:514–521, 1996.

[89] A. D. Spence and Y. Altintas. CAD Assisted Adaptive Control for Milling. *Trans. ASME Journal of Dynamic Systems, Measurement and Control*, 116:61–69, 1991.

[90] C. G. Goodwin and K. S. Sin. *Adaptive Filtering Prediction and Control*. Prentice-Hall, 1984.

[91] Y. Altintas. Direct Adaptive Control of End Milling Process. *International Journal of Machine Tool Design and Research*, 34(4):461–472, 1994.

[92] D. W. Clarke, C. Mohtadi, and P. S. Tuffs. Generalized Predictive Control – Part I. The Basic Algorithm. *Automatica*, 23(2):137–148, 1987.

[93] Y. Altintas, I. Yellowley, and J. Tlusty. The Detection of Tool Breakage in Milling Operations. *Trans. ASME Journal of Engineering for Industry*, 110:271–277, 1988.

[94] Y. Altintas. In-Process Detection of Tool Breakages Using Time Series Monitoring of Cutting Forces. *International Journal of Machine Tool and Manufacturing*, 28(2):157–172, 1988.

[95] Y. Altintas and P. Chan. In-Process Detection and Supression of Chatter in Milling. *International Journal of Machine Tool Design and Research*, 32:329–347, 1992.

[96] S. Smith and T. Delio. Sensor-Based Control for Chatter-Free Milling by Spindle Speed Selection. In *Proce. ASME 1989 WAM*, volume 18, pages 107–114, 1989.

[97] M. Weck, E. Verhagg, and M. Gather. Adaptive Control of Face Milling Operations with Strategies for Avoiding Chatter-Vibrations and for Automatic Cut Distribution. *Annals of the CIRP*, 24(1):405–409, 1975.

[98] C. K. Kuo. *Digital Control Systems,* 2nd ed. Saunders College Publishing, 1992.

INDEX